MACHINABILITY OF ENGINEERING MATERIALS

Machinability of Engineering Materials

B. MILLS

B.Sc., M.Sc., Ph.D., C.Eng., F.I.M., M.Inst.P.

and

A. H. REDFORD

B.Sc., Ph.D., A.R.T.C.S.

Department of Aeronautical and Mechanical Engineering, University of Salford, UK

APPLIED SCIENCE PUBLISHERS

LONDON and NEW YORK

APPLIED SCIENCE PUBLISHERS LTD
Ripple Road, Barking, Essex, England

Sole Distributor in the USA and Canada
ELSEVIER SCIENCE PUBLISHING CO., INC.
52 Vanderbilt Avenue, New York, NY 10017, USA

British Library Cataloguing in Publication Data

Mills, B.
Machinability of engineering materials.
1. Manufacturing processes
I. Title II. Redford, A.H.
670.42 TJ1185

ISBN 0-85334-183-4

WITH 96 ILLUSTRATIONS AND 22 TABLES

Photoset in Malta by Interprint Ltd
Printed in Great Britain by Galliard (Printers) Ltd, Great Yarmouth

PREFACE

In the manufacturing industries, despite the development and improvement of metal forming processes, a great deal of reliance is still placed on metal cutting processes and this will continue into the foreseeable future. Thus, there will continue to be a requirement for the development of improved cutting tool materials, workpiece materials, cutting fluids and testing methods; collectively this activity can be described as improving machinability.

Machinability is a parameter which in many ways is vague, sometimes qualitative and very often misunderstood. The purpose of this text is to give a broad understanding of the concept, methods of assessment and ways of improving machinability to the manufacturing engineer, the metallurgist and the materials scientist. The text should also be of interest to those engaged in research in manufacturing engineering and metal cutting. The text, of necessity, does not attempt to give detailed information about the machining characteristics of a wide range of tool and workpiece materials. It is felt that this is beyond the scope of the book and is best left to other sources, such as machinability data banks and the Machining Handbook*, whose main objective is to present this kind of information.

It is hoped that the reader will be able to progress logically from the fundamental aspects of the metal cutting process to the sections on the more specific topics of machinability including machinability testing and the properties of tool and workpiece materials which affect their machining performance. Finally, a section is devoted to the use of machinability data as an aid to 'optimising' the performance of metal cutting machines.

* *Metals handbook of machining, Vol. 3*, American Society for Metals (1967).

The major portion of the text is concerned with describing the work of other researchers. Whilst the present text cannot possibly reference all the work that has gone into metal cutting research we hope that the material presented is broadly representative of the work done.

During the last decade the authors have been fortunate in working and associating with many workers in the metal cutting field, in the United Kingdom, Europe, the United States of America and the Soviet Union. The number of such associates is too numerous to list; this would in any case be unwise for reasons of possible omissions. Our association with such colleagues and research students is greatly valued and appreciated.

B. MILLS AND A. H. REDFORD

CONTENTS

Chapter 1

THE CONCEPT OF MACHINABILITY

1.1. INTRODUCTION

At first sight, the definition of the term 'machinability' presents little difficulty. It is the property of a material which governs the ease or difficulty with which a material can be machined using a cutting tool. The term is in wide use by those concerned with engineering manufacture and production, yet detailed enquiries would reveal a measure of vagueness about its precise definition, or even its general meaning. Unlike most material properties, there is no generally accepted parameter used for its measurement and it is evident that, in practice, the meaning attributed to the term 'machinability' tends to reflect the immediate interests of the user. The engineer concerned especially with surface finish problems tends to think in terms of 'finishability', others may consider that the term can be used legitimately to indicate the consistency with which a material behaves in a particular machine tool set-up under a constant set of machining conditions, whilst some may consider it to be a measure of the useful life of the cutting tool. In most fields of science and technology great care is devoted to the definition of relevant parameters, but, in machining, machinability tends to remain a term which means 'all things to all men'.

1.2. DEFINITION OF MACHINABILITY

Attempts to arrive at an adequate definition of machinability tend to focus especially on one or more specific characteristics of the cutting process such as the cutting tool life, the tool wear rate, the energy required for a standard rate of metal removal, or the quality of the machined surface that can be produced. The cutting forces and the energy required for machining were, until recently, considered to be relatively unimportant for either operational or economic reasons in practice in industry; perhaps now, in an energy conscious environment, cutting efficiency could be considered to be a reasonable measure of machinability. A more general but still loose definition is based on the life of the cutting tool during which the quality of the workpiece is acceptable. Though it will be noted that, in many practical machining operations, surface quality is of no interest whatsoever, this definition means that in many circumstances, the life attributed to the cutting tool may depend simultaneously on two alternative characteristics: the tool wear, and the deterioration of surface quality. Furthermore, 'acceptable' surface quality and 'tool life' both need to be defined and this in itself presents considerable difficulties.

It therefore appears to be necessary to restrict deliberately, even for a general, and in many ways inadequate, definition, the meaning of the term 'machinability' or alternatively to coin a number of new terms each with a restricted meaning. As a practical step towards this goal, it is suggested that considerations of cutting energy and surface quality should not figure in the definition of machinability and that the term should be understood to be some measure of the way in which a material wears away a cutting tool when it is being machined.

Having restricted the meaning of machinability in this way, it is still necessary to specify how this characteristic of a material is to be measured. The ways in which the cutting tool may be worn are many, including, for example, flank wear, crater wear, notch wear, and edge chipping. Further, a number of possible parameters exist which could be used to describe each type of wear. There are also many different types of cutting tool with each major category differing in many respects from the others. Moreover, the conditions under which they may be used to machine material are infinitely variable. The choice of the type of wear to be measured, the parameter by which it is measured, the type of cutting tool, and the specific conditions under which it is to be used, all must be chosen somewhat arbitrarily for the definition and measurement of the machinability of the material. The machinability is thus a function of the test itself, and is clearly not a function of any one or two basic properties of the material. Moreover, since a fundamental understanding of the process of tool wear is lacking, it is not even possible to combine basic properties and cutting conditions to arrive at a measure of machinability. On the contrary, machinability must be specified empirically for a particular set of conditions, and this does not necessarily enable predictions to be made of the behaviour of the material when the conditions are changed. Although experimental work has shown that there may be some relationship between machinability observed in one type of test and that obtained in another test, or that measured under other conditions, such relationships remain almost entirely empirical. The rank order of machinability of a number of materials may change in different tests and under different conditions with the same test. Ultimately, the determination of the machinability of a material for a particular practical machining operation may have to be conducted under the specific conditions of that operation if a meaningful 'value' of the machinability is to be obtained.

These difficulties in defining and measuring the machinability of a material are paralleled in work concerned with determining the efficiency of cutting fluids, the wear resistance of cutting tool materials and the effectiveness of particular tool designs. Nevertheless, in spite of these fundamental difficulties, a considerable amount of work has been conducted in the search for a valid measure of machinability and in determining the behaviour of specific workpiece materials in practice. When reviewing this work, it is important that the basic limitations to the specification of machinability should be borne in mind.

Although machinability tests of many types have been conducted for a variety of metal removal operations, the bulk of machinability data has been generated for continuous-cutting operations and in particular for turning. Thus, whilst reference is made, where appropriate, to various aspects of machinability for a variety of processes, this text is concerned primarily with machinability in turning and the reader should avoid the temptation to translate recommendations for test procedures, tool materials and geometries, workpiece materials etc., into working propositions for other processes. Although some machinability features will be common to more than one process, given the same tool and workpiece materials, it is unfortunate that these cannot readily be identified. Thus it is likely that the short term savings of using 'equivalent' data and not carrying out machinability tests for the process under consideration will be more than offset by the high cost penalty which could result from the use of inappropriate cutting conditions.

Chapter 2

FUNDAMENTAL ASPECTS OF THE MACHINING PROCESS

2.1. MECHANICS OF METAL CUTTING

2.1.1. Chip Formation

The first explanations of the mechanics of metal cutting were formulated in the late 19th century. Early ideas proposed theories based on a 'splitting' of the workpiece material ahead of the cutting tool but this explanation was soon discarded in favour of the shear plane theory which suggests that the chip is formed during machining by fracture along successive shear planes which are inclined to the direction of cutting. A common and useful analogy which is made is that of a sliding pack of cards; from Fig. 2.1, it can be seen that the workpiece is

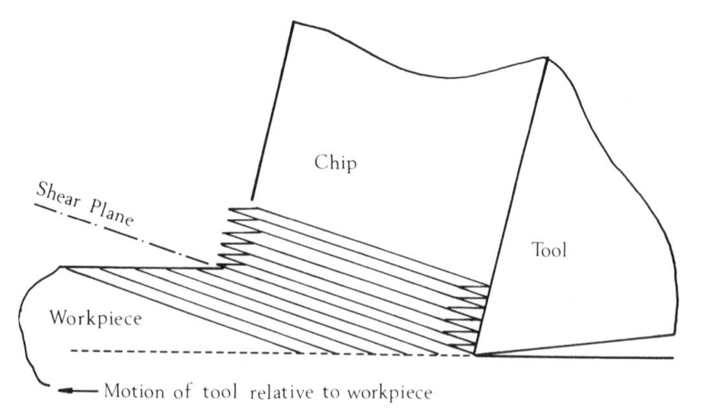

FIG. 2.1. Formation of a continuous chip.

considered to consist of a series of slender parallelograms which are inclined at some angle to the direction of relative work–tool motion. As the bottom right corner of an element reaches the cutting edge it will be deformed first elastically and then plastically until at some level of deformation the element will progressively slip relative to the following adjacent element until the bottom left corner of the element reaches the cutting edge. The plane along which the elements slip relative to each other is known as the shear plane and whilst in practice deformation takes place in a zone of finite width termed the 'shear' or 'primary deformation zone', most theories of metal cutting use the simple shear plane model as the basis for further calculation. More recently it has been suggested that shear in the shear zone is not continuous but that movement in this zone is effected by a series of very high frequency fractures and rewelds; little evidence exists to support this viewpoint.

When the newly formed chip attempts to move up the face of the cutting tool (Fig. 2.1), there is a very large normal force between the chip and the face and it has been suggested (1) that the distribution of normal stress on the face is as shown in Fig. 2.2. With this large stress, conditions are such that conventional coulomb friction does not apply where the real area of contact is directly proportional to the normal force and less than the apparent area of contact, and where the frictional force

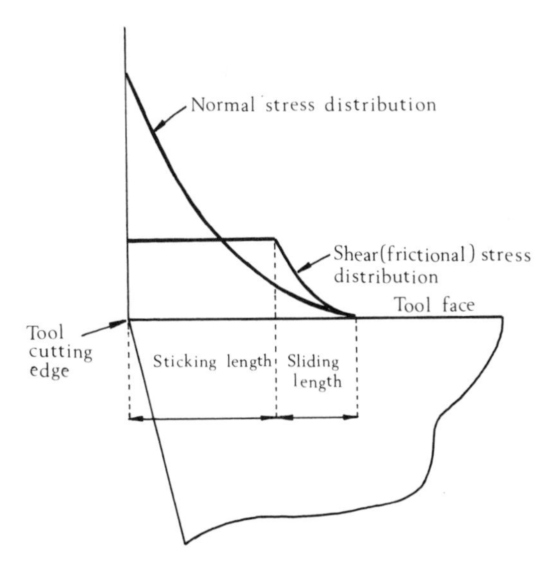

FIG. 2.2. Idealised model of tool face stress distribution (after Zorev (1)).

is proportional to the real area of contact. Instead, at the higher normal stress levels, the real area of contact is equal to the apparent area of contact and, under these conditions, increases in normal force have no effect on the real area of contact and the frictional force in this region becomes constant and equal to the force required to shear the chip material in this region, i.e. instead of the chip moving at uniform velocity up the face of the tool, the outer layers of the chip move at free chip velocity whilst the innermost layers have zero velocity relative to the tool. The distance up the face from the edge over which this phenomenon is evident is usually referred to as the 'sticking' length and from this position to that at which the chip loses contact with the face is usually referred to as the 'sliding' length. The existence of the sticking region has been verified by many workers and, in particular, conditions of seizure between tool and workpiece have been described in detail by Trent (2). A typical example showing the contact area between the tool and the underside of the chip is given in Fig. 2.3.

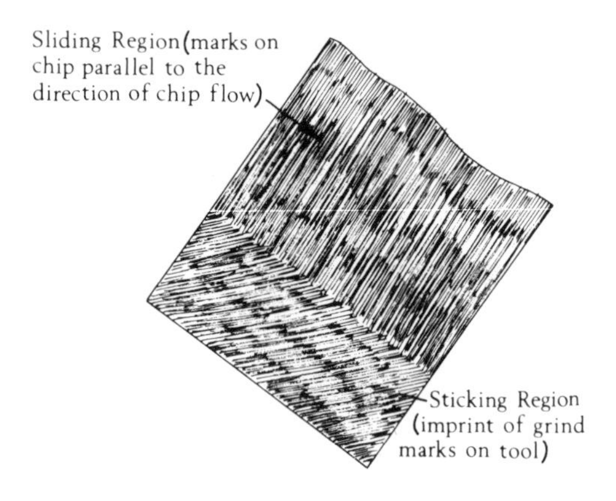

FIG. 2.3. Underside of chip showing regions of sticking and sliding friction.

Considerable deformation of the chip occurs over the sticking length (seizure zone) and this region is usually referred to as the secondary deformation zone. It has been suggested by many authors that this zone is nominally triangular, as shown in Fig. 2.4. Because there is always a sticking region in a metal cutting operation and because the thickness of the stationary material is finite, this zone is referred to as a built-up-layer (BUL). Under some cutting conditions the stationary layer is large and this is then referred to as a built-up-edge (BUE). Built-up-edges are of two types, unstable and stable. An unstable BUE periodically fractures

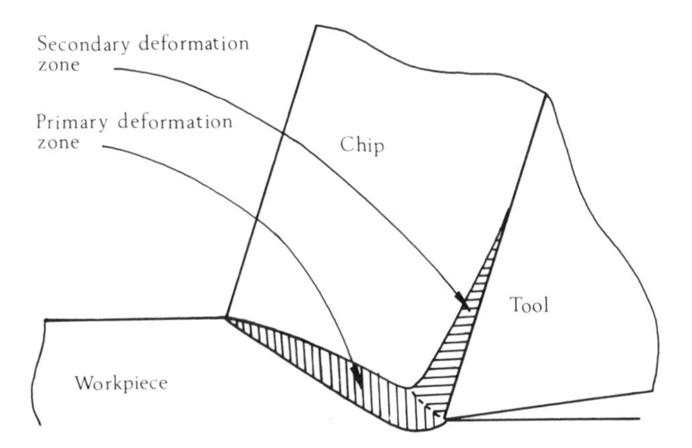

FIG. 2.4. Zones of deformation in orthogonal metal cutting.

and re-forms at high frequency and part of the edge is carried away by the underside of the chip and part passes beneath the flank of the tool and becomes embedded in the newly cut workpiece surface. Because the material which forms the BUE has been severely worked, it is usually much harder than the main body of the chip and this results in abrasive wear on both the face and the flank of the tool, and in a deterioration of the surface finish of the workpiece. Thus, an unstable BUE which occurs when machining ductile materials at low cutting speeds and/or under adverse machining conditions is invariably detrimental to both the life of the cutting tool and to the quality of the workpiece surface. Conversely, a stable BUE reaches a condition where the rate of formation of the BUE is equal to the rate of removal of the edge and, once formed—provided that steady state cutting is maintained—does not change in size. This type of BUE (Fig. 2.5) is invariably beneficial to the cutting process since, firstly, it increases the effective rake angle of the cutting tool which leads to a reduction in the cutting forces and, secondly, because the BUE protrudes beneath the flank of the tool, protection is given to the flank and this tends to decrease the amount of flank wear. Stable BUEs do not normally occur naturally but can be formed by the use of special cutting tool geometry. This will be discussed more comprehensively in Section 2.4 which is devoted to chip forming devices.

2.1.2. The Effect of Changes in Cutting Parameters on Cutting Forces

Whilst considerable effort has gone into producing theories to predict the results of metal cutting tests from a knowledge of the workpiece charac-

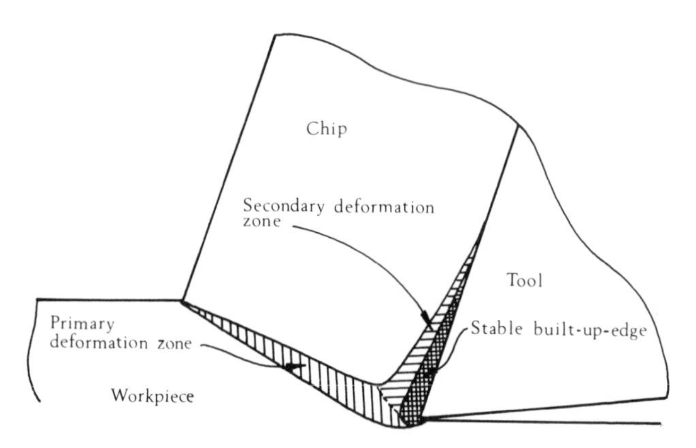

FIG. 2.5. Stable built-up-edge.

teristics, the geometry of the cutting tool and the cutting conditions, only limited success has been achieved. Even the most complex theories make assumptions which are often only true for a restricted set of conditions. Further, under the conditions of large strain and very large strain rate which apply in metal cutting, it is very difficult to obtain reliable data for the mechanical properties of materials under cutting conditions.

The major difficulty which arises when formulating a theory of metal cutting is imposed by the lack of constraint in the process which reduces the number of boundary conditions which can be applied. Recently it has been suggested (3) that for a particular set of cutting conditions no unique solution exists and that the values obtained for parameters such as cutting forces, cutting ratio and natural chip curl are dependent on the conditions which prevail when the tool first makes contact with the workpiece—if, for example, the initial cutting conditions are as shown in Fig. 2.6a then different answers will be obtained for steady cutting from those which would be obtained for steady cutting after initial conditions as shown in Fig. 2.6b.

Despite the lack of a comprehensive theory of metal cutting, general trends may be observed and these are now described.

Although, in practice, most metal cutting operations are oblique (Fig. 2.7a) in that the cutting edge is not perpendicular to the direction of relative work–tool motion and/or the direction of feed motion, it is convenient when considering the mechanics to examine an orthogonal cutting operation (Fig. 2.7b) since this is, effectively, a two-dimensional problem. For orthogonal cutting it is normal to measure forces parallel and perpendicular to the direction of relative work–tool motion and

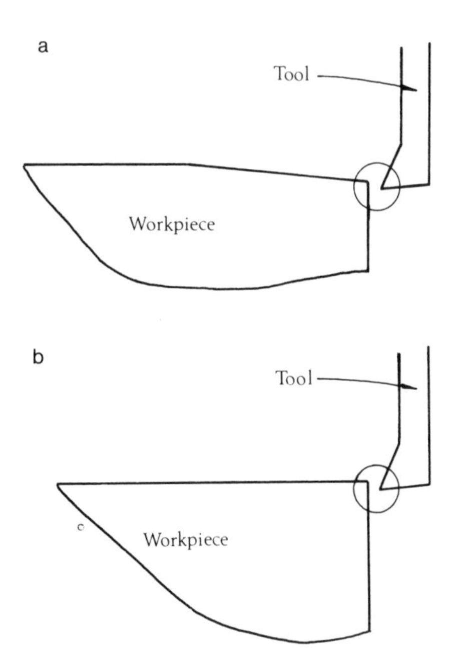

FIG. 2.6(a) Tapered entry machining; (b) Parallel entry machining.

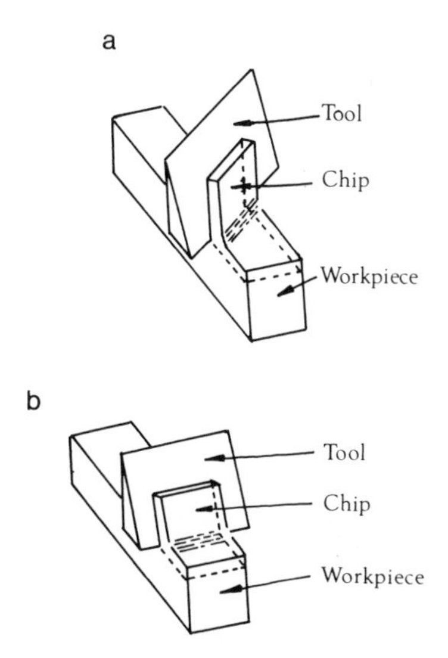

FIG. 2.7(a) Oblique cutting; (b) Orthogonal cutting.

these are defined as the tangential cutting force and the normal cutting force, respectively. The thickness of the chip prior to being sheared is termed the 'undeformed chip thickness', and the thickness of the chip after it has been sheared is termed the 'deformed chip thickness'. In an orthogonal operation the feed and the undeformed chip thickness are synonymous. The acute angle which the shear plane makes with the direction of relative work–tool motion is termed the 'shear angle', and the acute angle which the face makes with a plane drawn through the cutting edge perpendicular to the direction of relative work–tool motion is termed the 'rake angle'. These parameters are shown in Fig. 2.8. It is known that an increase in shear angle with a subsequent decrease in deformed chip thickness for a given set of cutting conditions is indicative of more effective cutting and results from decreasing the values of both tangential and normal cutting forces.

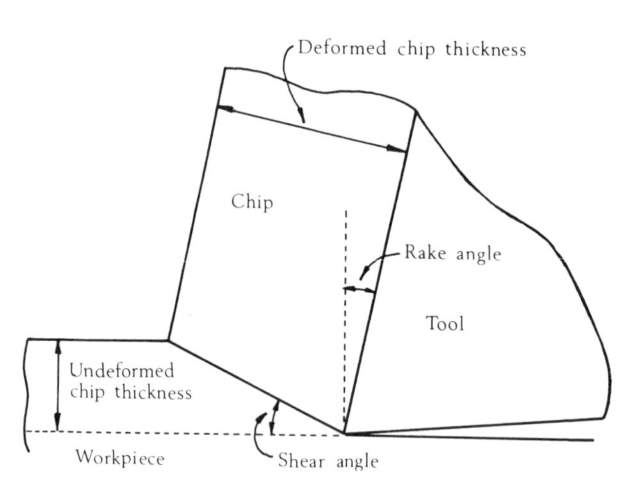

FIG. 2.8. Important parameters in orthogonal metal cutting.

For a plane face tool both tangential and normal cutting forces are directly proportional to the undeformed chip thickness and both have a finite value when the undeformed chip thickness is zero. This is a consequence of the 'nose' or 'ploughing' force which acts on the tool, i.e. the force which contributes nothing to metal removal and which is attributed to lack of sharpness of the edge however much care has been taken in the preparation of the tool. Thus, at small values of undeformed chip thickness, this force forms a significant proportion of the overall force acting on the tool, and cutting is relatively inefficient. As the undeformed chip thickness is increased, the ploughing force, which remains sensibly constant regardless of the value of undeformed chip thickness, becomes less significant and cutting becomes more efficient.

Increases in cutting speed in general reduce both the tangential and normal cutting forces but the effect is usually small. It is argued that as the cutting speed is increased, the temperature in both the primary and secondary deformation zones rises and this results in a 'softening' of the workpiece. This is a convenient, if simplified, explanation of a very complex process where both strain rate and temperature are increasing.

Increases in rake angle in the practical range always result in a decrease in both the tangential and normal cutting forces with, proportionately, a bigger decrease in the normal cutting force. This could be anticipated since a 'sharp' tool could be expected to cut more effectively than a 'blunt' tool.

Other changes in practical cutting geometries such as changes in plan approach angle, clearance angle, and tool corner radius have virtually no effect on the magnitude of the cutting forces with the exception of depth of cut for which increases in force are proportional to increases in depth.

2.1.3. The Effect of Changes in Cutting Parameters on Cutting Temperatures

In metal cutting operations heat is generated in the primary and secondary deformation zones and this results in a complex temperature distribution throughout the tool, workpiece and chip. A typical set of isotherms is shown in Fig. 2.9 where it can be seen that, as could be expected, there is a very large temperature gradient throughout the width of the chip as the workpiece material is sheared in primary deformation and there is a further large temperature gradient in the chip adjacent to the face as the chip is sheared in secondary deformation. This leads to a maximum cutting temperature a short distance up the face from the cutting edge and a small distance into the chip.

Since virtually all of the work done in metal cutting is converted into heat, it could be expected that factors which increase the power consumed per unit volume of metal removed will increase the cutting temperature. Thus an increase in the rake angle, all other parameters remaining constant, will reduce the power per unit volume of metal removed and the cutting temperatures will reduce. When considering increases in undeformed chip thickness and cutting speed the situation is more complex. An increase in undeformed chip thickness tends to be a

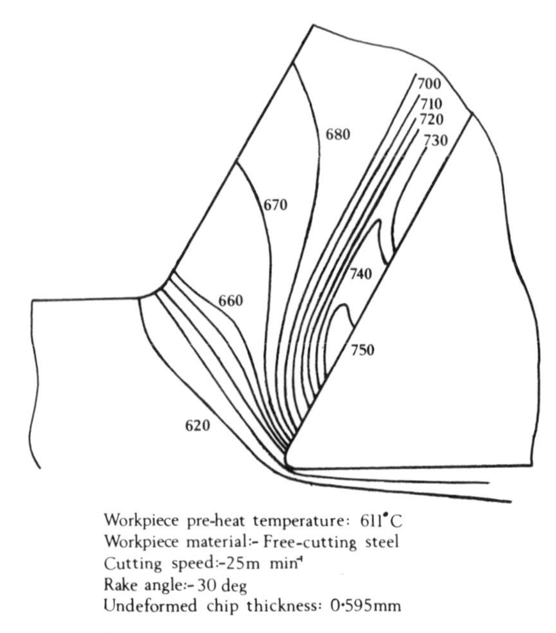

FIG. 2.9. Typical isotherms in the workpiece and the chip during orthogonal cutting.

scale effect where the amounts of heat which pass to the workpiece, the tool and the chip remain in fixed proportions and the changes in cutting temperature tend to be small. Increases in cutting speed, however, reduce the amount of heat which passes into the workpiece and this increases the temperature rise of the chip in primary deformation. Further, the secondary deformation zone tends to be smaller and this has the effect of increasing the temperatures in this zone. Other changes in cutting parameters have virtually no effect on the power consumed per unit volume of metal removed and consequently have virtually no effect on the cutting temperatures. Since it has been shown that even small changes in cutting temperature have a significant effect on tool wear rate it is appropriate to indicate how cutting temperatures can be assessed from cutting data. It has been found that predictions of cutting temperatures compare favourably with those obtained in practice and an analysis attributable to Weiner (4), Rapier (5) and Boothroyd (6) is presented in Appendix 1.

The most direct and accurate method for measuring temperatures in high-speed-steel cutting tools is that of Wright & Trent (7) which also yields detailed information on temperature distributions in high-speed-steel cutting tools. The technique is based on the metallographic examination of sectioned high-speed-steel cutting tools which relates microstructural changes to thermal history.

Trent (2) has described measurements of cutting tool temperatures and temperature distributions for high-speed-steel tools when machining a wide range of workpiece materials. This technique has been further developed by using scanning electron microscopy to study fine-scale microstructural changes arising from overtempering of the tempered martensitic matrix of various high-speed steels. This technique has also been used to study temperature distributions in both high-speed-steel single point turning tools (8) and twist drills (9).

2.2. TOOL WEAR

The wear processes of cutting tools which are relevant to cutting tool failure have been reviewed (10,11). The wear mechanisms include abrasive and adhesive wear, diffusion wear, wear arising from electrochemical action, and surface fatigue wear. Section 2.2.1 gives a brief summary of these wear mechanisms.

2.2.1. Mechanisms of Wear

2.2.1.1. Wear by Abrasion

The most common type of tool wear is that of abrasion where the relative motion between the underside of the chip and the face and the newly cut surface and the flank causes the tool to wear even though the newly cut workpiece surface and the chip may be very much softer than the tool material. In many cases, however, even though the workpiece and the chip may be relatively soft, hard inclusions or precipitates arising from the manufacturing process or from heat treatment will be present in the workpiece. Hard particles may also result from the breaking down of heavily work-hardened, unstable built-up-edges. Abrasive wear normally causes the development of a flat on the flank face and a crater on the face of the tool. Hard inclusions having sharp edges produce microcutting and give higher wear rates than hard, smooth, spherical inclusions which tend to groove the surface by plastic deformation rather than produce abrasive wear particles.

2.2.1.2. Wear by Adhesion

As has already been mentioned, pressure welding exists between the face of the tool and the underside of the chip under all cutting conditions. For those conditions where only a built-up-layer or a stable built-up-edge is present, although adhesion will occur, it will not result in the removal of tool material. However, when an unstable built-up-edge occurs, as well as particles of built-up-edge causing abrasive wear, it is likely that when the built-up-edge detaches itself from the face it will carry with it small quantities of tool material if strong bonding occurs between the built-up-edge and the tool material. Thus adhesive wear is primarily a wear mechanism on the face of the tool and usually occurs at low cutting speeds when an unstable built-up-edge is likely to be present. Trent (2) describes a form of this wear as attrition wear in which larger fragments of tool material of microscopic size are removed by adhesion.

2.2.1.3. Wear by Diffusion

Diffusion between cemented carbides and steel workpiece materials occurs at high cutting speeds and is a strongly temperature-dependent process in which atoms diffuse in the direction opposite to the concentration gradient (Fick's first law). Opitz & Konig (12) have shown that under the static conditions which occur in the seizure region on the face of a cutting tool, cobalt will diffuse into the steel. With the binding element removed a low shear strength layer exists on the surface of the tool which is transported from the tool by the underside of the chip. Trent (2) has shown that additions of titanium carbide (TiC) and tantalum carbide (TaC) reduce cratering wear by diffusion since they modify the structure of the tungsten carbide (WC) grains and this lowers their solubility in the workpiece.

2.2.1.4. Wear by Electrochemical Action

Under appropriate conditions, normally caused by the presence of a cutting fluid, it is possible to set up an electrochemical reaction between the tool and the workpiece which results in the formation of a weak low shear strength layer on the face of the tool. Whilst this is usually a desirable effect because it reduces the friction force acting on the tool which results in a reduction in the cutting forces and hence cutting temperatures, it will also usually result in small amounts of tool material being carried away by the chip. If the overall wear pattern is studied it is likely that the reduction in abrasive and, to some extent, adhesive wear which results from the action of the cutting fluid in reducing temperature and friction, respectively, will more than compensate for the small amounts of wear which occur due to electrochemical action.

In addition to the wear processes described above, tool material is sometimes removed by other mechanisms—the three most common being brittle fracture, edge chipping, and plastic deformation of the tool. Brittle fracture and edge chipping cause relatively large amounts of tool material to be removed whereas plastic deformation of the tool results in an adverse change in tool geometry which causes severe wear, usually on the tool flank.

Brittle fracture often causes a large portion of tool material to become detached from the tool—this results in instantaneous tool failure. This type of failure is normally associated with either extremely high forces acting on the tool due to the use of excessive feeds and/or depths of cut, or is due to the complex stress distribution set up in the tool under certain cutting conditions. If good metal cutting practice is adopted, the former should never result in failure since it should be possible to reduce the feed and/or depth of cut or to suitably strengthen the tool. The latter is normally associated with the cutting of high strength materials with carbide cutting tools and it has been shown (13) that as the flank wear land starts to develop the stress pattern in the tool is modified until, even with a relatively small flank wear land, tensile stresses are set up within the tool. Since the tool is weak in tension this will often result in tool failure.

Edge chipping is a common wear phenomenon in intermittent cutting operations where cyclic mechanical and thermal stresses are applied to the tool; this results in fine cracks developing near to the cutting edge and flaking of tool material. Plastic deformation of the cutting edge, particularly the tool corner, is caused by high temperatures and stresses and is therefore primarily a high cutting speed effect in which high tool temperatures are generated.

2.2.1.5. Wear by Fatigue

Fatigue wear is only an important wear mechanism when adhesive and abrasive wear rates are small. Surfaces which are repeatedly subjected to loading and unloading may gradually fail by fatigue leading to detachment of portions of the surface. This situation can arise in intermittent cutting which may also cause edge chipping. Nucleation of subsurface fatigue cracks may be initiated at subsurface defects such as non-metallic inclusions. Fatigue cracking does not normally occur if the stress is below a certain limit. Since the contact pressures are determined by the yield properties of the workpiece material, fatigue wear can be reduced by using cutting tools which are appreciably harder than the workpiece.

2.2.2. Types of Wear

Discounting brittle fracture and edge chipping, which have already been dealt with, tool wear is basically of three types. Flank wear, crater wear, and notch wear. Flank wear occurs on both the major and the minor cutting edges. On the major cutting edge (Fig. 2.10) which is responsible

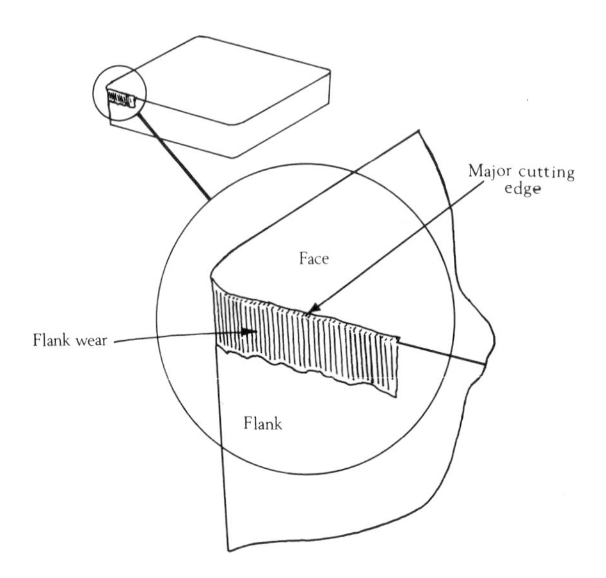

FIG. 2.10. Flank wear on an indexable insert.

for bulk metal removal, this results in increased cutting forces and higher temperatures which if left unchecked can lead to vibration of the tool and workpiece and a condition where efficient cutting can no longer take place. On the minor cutting edge which determines workpiece size and surface finish, flank wear can result in an oversized product which has poor surface finish. Under most practical cutting conditions, the tool will 'fail' due to major flank wear before the minor flank wear is sufficiently large to result in the manufacture of an unacceptable component.

Because of the stress distribution on the tool face, the frictional stress in the region of sliding contact between the chip and the face is at a maximum at the start of the sliding contact region and is zero at the end. Thus abrasive wear takes place in this region with more wear taking place adjacent to the seizure region than adjacent to the point at which the chip loses contact with the face. This results in localised pitting of the tool face some distance up the face which is usually referred to as cratering and which normally has a section in the form of a circular arc (Fig. 2.11). In many respects and for practical cutting conditions, crater

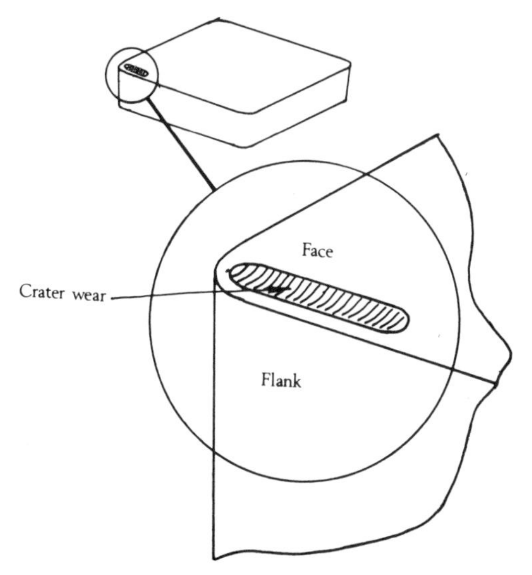

FIG. 2.11. Crater wear on an indexable insert.

wear is a less severe form of wear than flank wear and consequently flank wear is a more common tool failure criterion. However, since various authors have shown that the temperature on the face increases more rapidly with increasing cutting speed than the temperature on the flank, and since the rate of wear of any type is significantly affected by changes in temperature, crater wear usually occurs at high cutting speeds.

At the end of the major flank wear land where the tool is in contact with the uncut workpiece surface it is common for the flank wear to be more pronounced than along the rest of the wear land. This is because of localised effects such as a hardened layer on the uncut surface caused by work hardening introduced by a previous cut, an oxide scale, and localised high temperatures resulting from the edge effect. This localised wear is usually referred to as notch wear (Fig. 2.12) and occasionally is very severe. Although the presence of the notch will not significantly

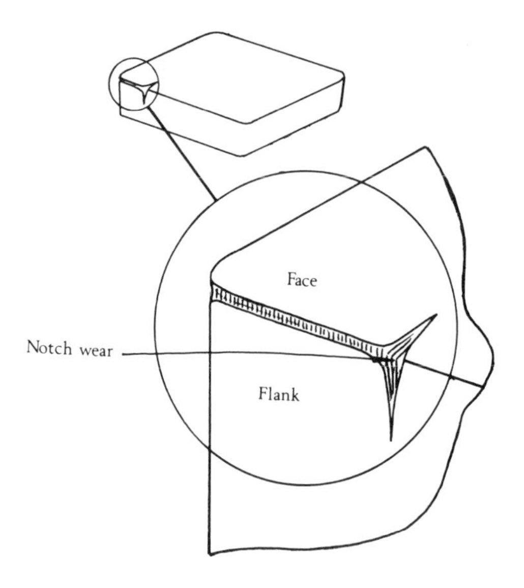

FIG. 2.12. Notch wear on an indexable insert.

affect the cutting properties of the tool, the notch is often relatively deep and if cutting were to continue there would be a good chance that the tool would fracture.

If any form of progressive wear were allowed to continue, eventually the wear rate would increase dramatically and the tool would fail catastrophically, i.e. the tool would be no longer capable of cutting and, at best, the workpiece would be scrapped whilst, at worst, damage could be caused to the machine tool. For carbide cutting tools and for all types of wear, the tool is said to have reached the end of its useful life long before the onset of catastrophic failure. For high-speed-steel cutting tools, however, where the wear tends to be non-uniform it has been found that the most meaningful and reproducible results can be obtained when the wear is allowed to continue to the onset of catastrophic failure even though, of course, in practice a cutting time far less than that to failure would be used. The onset of catastrophic failure is characterised by one of several phenomena, the most common being a sudden increase in cutting force, the presence of burnished rings on the workpiece, and a significant increase in the noise level.

2.2.3. Relationship Between Tool Wear and Time

For progressive flank wear the relationship between tool wear and time follows a fixed pattern. Initially, with a new tool, the tool wear rate is high and is referred to as primary wear. The time for which this wear rate acts is dependent on the cutting conditions but, typically, for a given workpiece material, the amount of primary wear is approximately constant but the time to produce it decreases as the cutting speed is increased. This wear stage is followed by the secondary wear stage where the rate of increase of flank wear is sensibly constant but considerably less than the rate of primary wear in the practical cutting speed range. At the end of the secondary wear stage, when the flank wear is usually considerable and far greater than that recommended as the criterion for tool failure, the conditions are such that a second rapid wear rate phase commences (tertiary wear) and this, if continued, rapidly leads to tool failure. The three stages of wear are illustrated in Fig. 2.13. It is often suggested that the high rate of wear in the primary wear stage is due to edge crumbling and is not typical of a 'worn-in' tool. However, it has been suggested (14) that it is not the primary wear rate which is large for the tool–workpiece combination but that the reduced secondary wear rate is a consequence of the protection afforded to the tool by the small stable built-up-edge which forms as the edge is removed from the tool.

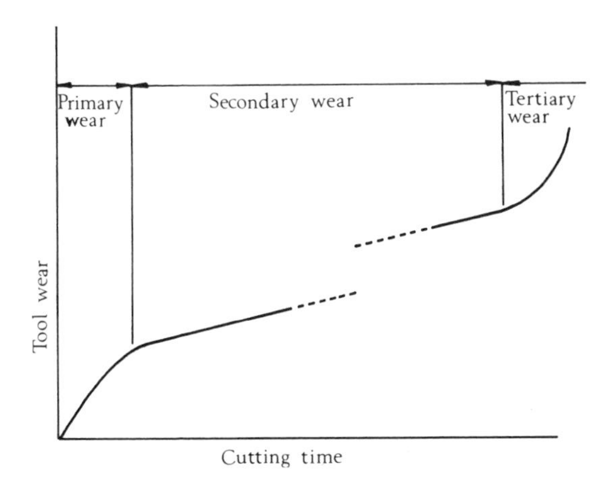

FIG. 2.13. Typical relationship between flank wear and cutting time.

Crater wear, normally measured in terms of the depth of the crater, increases progressively with time until a point is reached when the crater weakens the tool sufficiently for the forces acting on the tool to fracture it. Thus the criterion for tool failure due to crater wear is based on a crater depth of a constant amount plus a further amount which is proportional to the feed. Catastrophic failure of high-speed-steel tools is merely an extension of the flank wear criterion for carbides and follows the same type of relationship with time. All other forms of wear which result in rapid deterioration of the tool are often difficult to relate to time in a meaningful manner since the tool can fail when there is little or no wear and this can often be due to a transient condition in what is basically a steady-state operation.

2.2.4. Relationship Between Tool Wear and Cutting Conditions

F. W. Taylor in his classic paper '*On the art of cutting metals*' (15) suggested that, for progressive wear, the relationship between the time to tool failure for a given wear criterion and cutting speed was of the form

$$VT^{-1/k} = C_1 \qquad [2.1]$$

where V is the cutting speed; T is the tool life; and k and C_1 are constants for the tool–workpiece combination.

This basic relationship has been tested repeatedly for a wide range of tool and workpiece materials and cutting conditions and, except at very low or very high cutting speeds and provided the tool failure criterion does not change, has been found to be valid. The equation was later extended to the more general form

$$T = \frac{C_2}{V^p S^q a^r} \qquad [2.2]$$

where S is the feed; a is the depth of cut; and p,q,r and C_2 are constants for the tool–workpiece combination.

In 1932, Woxen (16) suggested a method of combining the effects of feed, depth of cut, and tool geometry using what he termed 'the chip equivalent' concept. Recently, further work has been done on this aspect of machinability and this will be dealt with in Chapter 3 which is devoted to short-term testing.

Considering the major variables of speed, feed, and depth of cut, in general, by far the most significant is cutting speed where for modern carbide cutting tools p will be of the order of 2 to 4 and for high-speed-steel cutting tools will lie between 4 and 7. In contrast to this, q will usually be unity or less and r, the constant associated with depth of cut, will often be very small and negative, i.e. as the depth of cut is increased, the tool life tends to increase slightly.

2.2.5. Relationship Between Tool Life and Temperature

It has been known for many years that as the tool temperature increases, the tool life reduces; the relationship between the tool life and temperature is of the form

$$\theta T^{n} = C_3 \qquad [2.3]$$

where θ is some measure of tool temperature; and n and C_3 are constants for the tool–workpiece combination.

The temperature can be measured in a variety of ways but the most common method uses a work–tool thermocouple, i.e. a device which uses the dissimilar material junction between the tool and the workpiece as a means of generating an e.m.f. which is proportional to the temperature of the junction; but, since the junction temperature will vary considerably from place to place along the junction it is difficult to say exactly what is being measured. However, it has been found that the temperature as recorded by a work–tool thermocouple when used to plot a θ–T relationship gives good results. Typically, the exponent n in this relationship is between 0·05 and 0·1 and this indicates how critically cutting temperature affects tool life.

2.2.6. Tool Life Criteria

Recently, the International Standards Organisation (ISO) suggested a standard for tool life testing (ISO 3685-1977)—this is discussed in detail in Chapter 6. Embodied in the standard are tool life criteria for both high-speed-steel and carbide cutting tools where, with the exception of catastrophic failure, all types of progressive wear have specified amounts of wear for which the tool is considered to have failed.

2.3. SURFACE FINISH

2.3.1. Introduction

Surface finish is a complex parameter which is very difficult to specify in a meaningful way. Many methods have been developed to measure surface finish from the very simple visual or touch comparator methods where a surface is compared with one of a series of standard surfaces to methods using sophisticated equipment which will measure the root mean square (RMS) average of a surface or the centre line average (CLA) of a surface. Probably the most accepted measure of surface finish is the

CLA value but two main problems exist when measuring this or any other similar parameter. Firstly, it is quite possible that two surfaces with an identical CLA value can have completely different basic topographies. In many cases this will not be particularly important but for mating moving surfaces, for example, where oil retention may be important, any measure of surface finish which is available will not fully describe what is required. Secondly, in many metal removal operations, surface finish properties are polarised in that the value obtained will depend on the direction of measurement. Perhaps the best example of this is turning where the surface finish in the direction of tool feed motion is usually poorer than that in a direction perpendicular to tool feed motion. For most metal cutting operations the specification is for surface finish to be measured in a given direction and this will usually be the direction in which it is expected that the surface finish will be the poorest.

2.3.2. Mechanism of Surface Finish Production

There are basically five mechanisms which contribute to the production of a surface which has been machined. These are:

(1) The basic geometry of the cutting process. In, for example, single point turning the tool will advance a constant distance axially per revolution of the workpiece and the resultant surface will have on it, when viewed perpendicularly to the direction of tool feed motion, a series of cusps which will have a basic form which replicates the shape of the tool in cut. Figure 2.14 shows an example of a surface which has been produced using a tool having a corner radius and it can be shown that for this shape of tool the

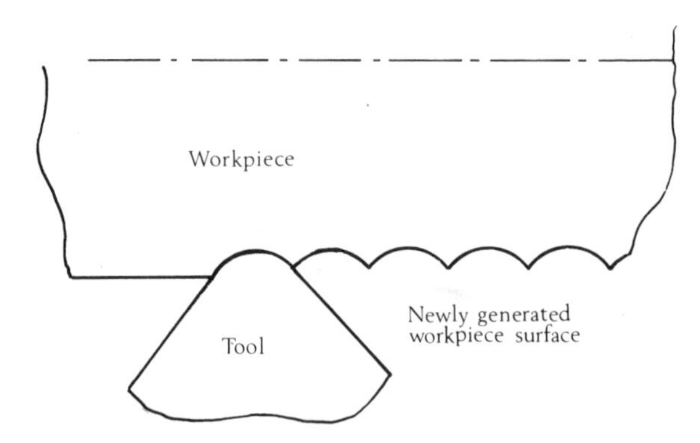

FIG. 2.14. Workpiece surface generated using a tool having a corner radius.

geometric CLA of the surface is given by

$$32{\cdot}1(S^2/R) \qquad \text{(microns)} \qquad [2.4]$$

where S is the feed rate (mm min^{-1}); and R is the tool nose radius (mm).

(2) The efficiency of the cutting operation. It has already been mentioned that cutting with unstable built-up-edges will produce a surface which contains hard built-up-edge fragments which will result in a degradation of the surface finish. It can also be demonstrated that when cutting under adverse conditions such as apply when using large feeds, small rake angles and low cutting speeds, besides producing conditions which lead to unstable built-up-edge production, the cutting process itself can become unstable and instead of continuous shear occurring in the shear zone, tearing takes place, discontinuous chips of uneven thickness are produced, and the resultant surface is poor. This situation is particularly noticeable when machining very ductile materials such as copper and aluminium.

(3) The stability of the machine tool. Under some combinations of cutting conditions, workpiece size, method of clamping, and cutting tool rigidity relative to the machine tool structure, instability can be set up in the tool which causes it to vibrate. Under some conditions this vibration will reach and maintain a steady amplitude whilst under other conditions the vibration will build up and unless cutting is stopped considerable damage to both the cutting tool and workpiece may occur. This phenomenon is known as chatter and in axial turning is characterised by long pitch helical bands on the workpiece surface and short pitch undulations on the transient machined surface.

(4) The effectiveness of removing swarf. In discontinuous chip production machining, such as milling or turning of brittle materials, it is expected that the chips (swarf) will leave the cutting zone either under gravity or with the assistance of a jet of cutting fluid and that they will not influence the cut surface in any way. However, when continuous chip production is evident, unless steps are taken to control the swarf it is likely that it will impinge on the cut surface and mark it. Inevitably, this marking, besides looking unattractive, often results in a poorer surface finish.

(5) The effective clearance angle on the cutting tool. For certain geometries of minor cutting edge relief and clearance angles it is

possible to cut on the major cutting edge and burnish on the minor cutting edge. This can produce a good surface finish but, of course, it is strictly a combination of metal cutting and metal forming and is not to be recommended as a practical cutting method. However, due to cutting tool wear, these conditions occasionally arise and lead to a marked change in the surface characteristics.

2.3.3. Factors Which Influence Surface Finish Production

2.3.3.1.

The basic geometry of the cut surface in influenced primarily by the tool geometry and the feed. It is unusual for this contributor to surface finish to present any technical problems in practice but in general it could be said that there is a cost penalty, in time, which has to be paid for improving the quality of the geometric surface. This condition only applies for a given process; clearly some cutting processes inherently produce a better surface finish per unit cost than others.

2.3.3.2.

Under normal cutting conditions, when cutting most materials, unstable built-up-edge production will not usually present a severe problem and the effect of built-up-edge fragments on the workpiece will be small particularly if carbide or ceramic cutting tools are used at economic cutting speeds. Thus, in practice, degradation of the surface from the geometric surface due to adverse cutting conditions is caused by factors which can be controlled and it should therefore be possible to eliminate most of the problems. If the cutting speed can be set high then the adverse effect of small tool rake angles becomes much less critical and, as a consequence, for practical rake angles and common ferrous workpieces it would be unusual to produce significant extra surface roughness when cutting at high speed. However, when cutting ductile materials, even at high speed, the choice of rake angle is very important and, from the surface finish aspect only, increasing the rake angle tends to improve the machining conditions and improve surface finish. Even when cutting at high speed, however, many non-ferrous ductile materials produce conditions where an irregular and often unstable built-up-edge is formed and this can have a marked adverse effect on surface finish. Invariably, the only way that a good surface finish can be produced when machining these materials is by using a cutting fluid which will prevent built-up-edge formation.

A badly adjusted obstruction-type chip former or a poor geometry groove-type chip former will often lead to a poorer surface finish if cutting results in severely 'overbroken chips', i.e. chips which are too tightly curled. To maintain the cutting conditions would require that the obstruction be moved further away from the cutting edge in the case of an obstruction-type former or that the groove width be increased for a groove-type chip former. If neither of these actions is possible, then a similar effect could be achieved by reducing the feed.

2.3.3.3.
Machine tool vibrations—particularly the phenomenon of chatter—have been thoroughly investigated in the past yet, unfortunately, the methods by which chatter is eliminated are still often not predictable. Clearly, increasing or decreasing the stiffness of the tool mounting structure will, for a given severe chatter condition, tend to reduce the effect and usually it would be appropriate to stiffen the tool mounting structure. In a particular situation where, within reason, the stiffness of the tool mounting structure is fixed, other solutions have to be found. One possible solution is to increase the stiffness of the workpiece by utilising a better clamping arrangement, e.g. if, in turning, a chuck-mounted workpiece is chattering, it may be possible to reduce the overhang of the bar, mount the bar between centres, mount the bar between chuck and centres or use a fixed or travelling steady.

If the workpiece geometry and clamping are fixed then changes in cutting condition will be necessary and it is most common firstly to investigate the effect of changes in cutting speed. If these changes do not produce the desired effect then a change in feed may be beneficial, particularly an increase in feed; unfortunately, of course, this action would also produce a rougher geometric surface. A further alternative which can have a beneficial effect is to use or change the cutting fluid.

2.3.3.4.
Swarf control is invariably achieved by using chip control devices and this will be dealt with in more detail in Section 2.4.

2.4. CHIP FORMERS

2.4.1. Mechanics of Chip Formers

In continuous chip production metal cutting operations, swarf control is necessary for the following reasons:

(1) Continuous chips have a very low effective density and present problems when swarf removal is attempted either manually or automatically.
(2) Continuous chips can wrap themselves around the workpiece and whilst this will generally only have a small effect on the surface finish, the scuffing leads to an unattractive surface.
(3) When continuous chips wrap themselves around the workpiece they present a safety hazard to operators on manually controlled machines.

Many methods of effectively breaking continuous chips have been proposed but by far the most successful have resulted from the use of chip control devices. These may be categorised into two broad groups: the obstruction-type, and the groove-type chip formers. The first are of two types, step formers and ramp formers, whilst the second may be of the simple single groove type of various forms or more recently of the multigroove type. The multigroove type of chip former has been developed to allow one chip former to perform effectively over a wide range of cutting conditions, in particular for a reasonable range of changes in feed.

Whatever the type of chip former, the basic action is the same. The chip is constrained to move in a circular path and under practical cutting conditions this results in the free end of the chip impinging on the flank face of the cutting tool or tool-holder. At this stage the free end of the chip anchors itself to the contact point and subsequent chip production results in an increase in the radius of curvature of the formed chip. Under favourable circumstances, as the chip and the radius increase, a stage is reached when the stresses in the chip are such that it breaks. Figure 2.15 shows a chip on the point of breaking and it can be seen that the maximum stress occurs approximately half way between the contact point of the free end of the chip and the point at which the chip leaves the face of the tool.

Using a chip forming device, the chips produced may be classified as one of three types. Firstly, if the initial chip flow circle radius is too large and/or the thickness of the chip is too small, the chip will be underbroken, i.e. there is a strong possibility that, as the chip expands, the stresses in it will not be sufficient to cause it to fracture before it becomes unstable and the free end detaches from its anchoring point. Under these circumstances chip breaking will not be achieved and large radius, continuous helical chips will result. Secondly, if the chip flow circle radius and the thickness of the chip are appropriate, the chip will break

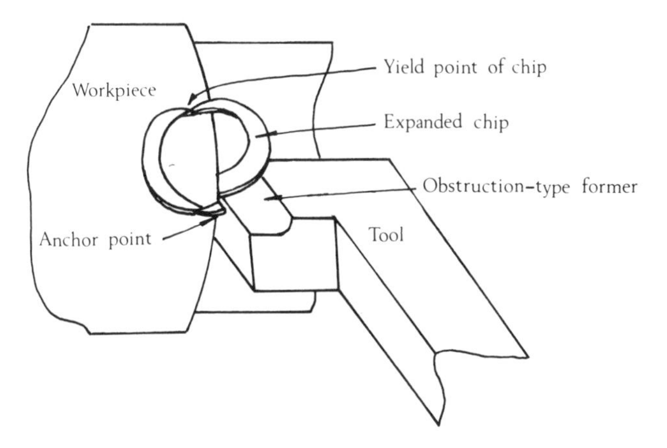

FIG. 2.15. Effectively broken chip at the point of breaking.

consistently on expansion and ideally the broken chips will be of the half-moon type where, typically, the overall length of the chip is of the order of 20 mm and the chip curl radius is of the order of 5 mm. Under these circumstances, damage to the workpiece surface is minimal and the swarf is relatively compact and easily disposable by either manual or automatic means. Thirdly, if the chip curl radius is small and/or the thickness of the chip is large, the chip will be overbroken, i.e. the chip breaks into small, tightly curled segments which fly from the cutting zone at high velocity. It is unlikely that this type will cause damage to the surface directly but it is possible that adverse conditions will be encountered and this can result in surface damage. Figure 2.16 shows examples of typical underbroken, effectively broken, and overbroken chips.

2.4.2. Effect of Cutting Conditions on Chip Forming

Of the main cutting parameters, the two most significant with regard to chip forming are the feed and the plan approach angle of the cutting tool. Considering the latter first, it has been shown (17) that the chip flow angle is equal to the plan approach angle and thus, if the plan approach angle is too small, the helix angle of the chip will not be sufficient for the free end of the curved chip to miss the uncut workpiece surface. Under these circumstances it is still possible for the chip to be carried round by the workpiece and for it to anchor on the flank of the tool or tool-holder but this is unpredictable. If the plan approach angle is too large then it is likely that the large helix angle chip will miss the flank of the tool

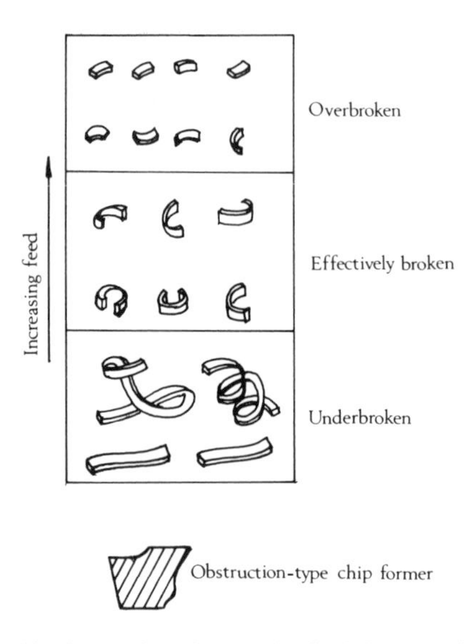

FIG. 2.16. Typical chip forms for changes in feed for an obstruction-type chip former.

completely and breaking will not take place. In practice, a standard plan approach angle of 15° will often yield consistent results. For an obstruction-type chip former of fixed geometry, as the feed is increased, the chip progressively passes through the underbroken, effectively broken and overbroken stages because the chip flow circle radius reduces and the chip is thicker. A typical graphical representation of this, attributable to Henriksen (18), is shown in Fig. 2.17.

An increase in the cutting speed tends to increase the radius of the chip flow circle because it reduces the contact length between the tool face and the underside of the chip. Also, with increased cutting speed the chip is often more ductile and thus a speed increase will tend to cause an effectively broken chip to underbreak. In order to maintain effective breaking, therefore, an increase in cutting speed would have to be accompanied by an increase in feed. An increase in tool rake angle also decreases the contact length between the chip and the tool and this has a similar effect to an increase in speed. As for increases in speed, the solution to maintaining the same chip breaking condition is to increase the feed.

An increase in depth of cut will not significantly affect the chip flow circle radius but, because of the width of the chip, it may be necessary to

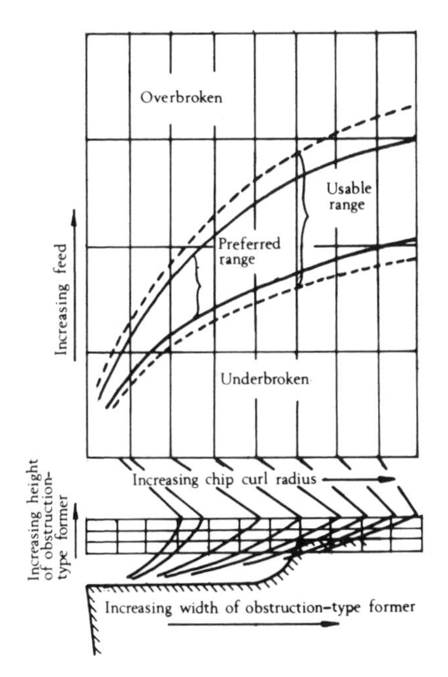

FIG. 2.17. Graphical representation of chip breaking using an obstruction-type chip former (after Henriksen (18)).

increase the plan approach angle of the tool to ensure that the formed chip does not impinge on the uncut workpiece surface.

For simple groove-type chip formers operating in the practical range an increase in feed also causes a decrease in chip flow circle radius to a minimum which is normally imposed by the radius of the groove. Thus an increase in feed will cause the chip to change from underbroken to effectively broken to overbroken. A condition which has to be noted for this type of device is that there is a minimum feed which must be used before chip forming by the groove becomes effective. This value of feed is usually referred to as the 'straightening' feed at which condition the contact length between the tool and the chip is equal to the land length of the chip former and the chip is formed straight (Fig. 2.18). Increases in cutting speed, rake angle, and depth of cut will have the same type of effect as found for an obstruction-type former.

2.4.3. Effect of Chip Formers on Cutting Forces

It has been shown by many previous researchers that the forces which act on the forming elements of chip formers are very small in comparison

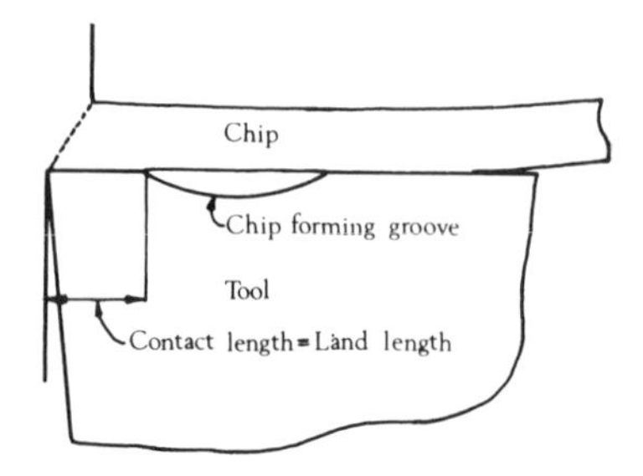

FIG. 2.18. Cutting conditions for chip straightening.

with the forces required to produce the chip in the primary and secondary deformation zones. It has also been shown that the effect of the former is to modify the conditions which exist in the primary deformation zone such that the chip is formed curled. Thus, for obstruction-type chip formers, the presence of the former, particularly if it is working under effective chip breaking conditions, will have very little effect on the cutting forces when compared with a plane faced tool working under identical cutting conditions.

For groove-type forming, however, the cutting process is significantly different in that when the former is working effectively a large stable built-up-edge forms on the land of the chip former and the angle of this built-up-edge is equal to the entry angle of the chip forming groove, as shown in Fig. 2.19. Thus the chip is formed along the built-up-edge and the effective rake angle of the tool becomes the nominal rake angle plus the built-up-edge angle. With this increase in rake angle the cutting forces reduce and, as has been discussed in Section 2.1.3, this will result in a decrease in cutting temperature.

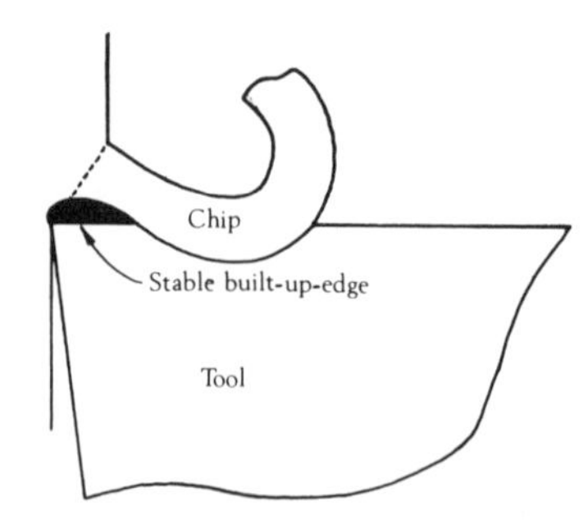

FIG. 2.19. Groove-type chip forming showing large stable built-up-edge.

2.4.4. Effect of Chip Formers on Tool Wear

Tests on various types of chip former have indicated that, when compared with a plane faced tool, a tool with an obstruction-type chip former does not perform significantly differently with regard to tool flank wear whereas for a groove-type chip former the evidence suggests that the flank wear rate is slightly reduced. However, these results were not conclusive in that the amount of change was not sufficient to be sure that the observed trend could be fully justified statistically. This was surprising because it was thought that the combination of a large stable built-up-edge and the reduction in cutting temperature would have resulted in a significantly reduced wear rate.

2.5. THE ACTION OF CUTTING FLUIDS

Cutting fluids are of two basic types, cutting lubricants and soluble oils (coolants). As the name implies, a cutting lubricant, if effective, reduces friction on the face of the tool which results in a decrease in the cutting forces and cutting temperatures. It has been suggested by some authors that some lubricants can reduce the shear stress in the primary shear zone but this has not been substantiated.

In contrast to a lubricant and, again, as its alternative name implies, the main function of a soluble oil is to cool. Whilst some reduction in the cutting temperature in the shear zones could be expected when using a coolant, its main purpose is to keep the workpiece temperature stabilised and as little above ambient temperature as possible so that, particularly for close tolerance work, the job is not under size when it cools on completion of cutting.

For continuous-cutting operations when there is continuous contact between the face and the chip it is difficult to introduce a cutting lubricant to the face and, with the exception of low speed cutting with 'special' lubricants (19), that lubricating action is very limited. For discontinuous-cutting operations, however, particularly when the duration of the cut is small, there is an opportunity to apply the lubricant to the face of the tool every time it comes out of cut and under these circumstances genuine lubrication can take place.

Conversely, whilst a coolant is usually successful in doing its job in continuous-cutting operations, there is a possibility that in intermittent cutting the tool will be subjected to thermal shock which in turn may lead to breakage of the tool.

In virtually all cases, the use of both cutting lubricants and coolants results in a decrease in cutting temperature and an increase in tool life. Further, both tend to improve the efficiency of the cutting process and this tends to lead to an improvement in surface finish.

REFERENCES

1. ZOREV, N. N. (1963). International Research in Production Engineering, *International Production Engineering Research Conference*, Pittsburgh, 142.
2. TRENT, E. M. (1977). *Metal cutting*, Butterworths, London.
3. DEWHURST, P. (1978). On the non-uniqueness of the machining process, *Proc. R. Soc. Lond.*, **A 360**, 587–610.
4. WEINER, J. H. (1955). Shear plane temperature distribution in orthogonal cutting, *Trans. A.S.M.E.*, **77**, 1331.
5. RAPIER, A. C. (1954). A theoretical investigation of the temperature distribution in orthogonal cutting, *Brit. J. Appl. Phys.*, **5**, 400.
6. BOOTHROYD, G. (1963). Temperatures in orthogonal metal cutting, *Proc. Inst. Mech. Engrs.*, **177**, 789.
7. WRIGHT, P. K. & TRENT, E. M. (1973). Metallographic methods of determining temperature gradients in cutting tools, *J.I.S.I.*, **211**, 364.
8. MILLS, B., WAKEMAN, D. W. & ABOUKHUSHABA, A. (1980). A new technique for determining the temperature distribution in high-speed steel cutting tools using scanning electron microscopy, *Annals of C.I.R.P.*, **29**, 73.
9. MILLS, B. & MOTTISHAW, T. D. (1981). The application of scanning electron microscopy to the study of temperatures and temperature distributions in M2 high-speed steel twist drills, *Annals of C.I.R.P.*, **30**, 15.
10. TABOR, D. (1970). Some basic mechanisms of wear that may be relevant to tool wear and tool failure, *ISI Report*, **126**, 16.
11. TRENT, E. M. (1970). Wear processes which control the life of cemented carbide cutting tools, *ISI Report*, **126**, 10.
12. OPITZ, H. & KONIG, W. (1967). On the wear of cutting tools, *8th Int. M.T.D.R. Conference*, 173.
13. ELLIS, J. & BARROW, G. (1972). The failure of carbide tools when machining high strength steels, *Annals of C.I.R.P.*, **21**, 25.
14. REDFORD, A. H. (1980). The effect on cutting tool wear of various types of chip control device, *Annals of C.I.R.P.*, **29**, 67.
15. TAYLOR, F. W. (1907). On the art of cutting metals, *Trans, A.S.M.E.*, **28**, 31–432.
16. WOXEN, R. (1932). A theory and equations for the life of lathe tools, *Ingeniors Vetenskaps, Akademien Handlingar*, **119**, 73.
17. STABLER, G. V. (1951). The fundamental geometry of cutting tools, *Proc. Inst. Mech. Engrs.*, **14**, 165.
18. HENRIKSEN, E. K. *A study of the chip breaker*, A.S.M.E. Paper No. 53-S-9.
19. CASSIN, C. & BOOTHROYD, G. (1965). Lubricating action of cutting fluids, *J. Mech. Eng.*, **7**(1), 67–81.

Chapter 3

THE ASSESSMENT OF MACHINABILITY

3.1. TYPES OF MACHINABILITY TEST

Machinability tests may be subdivided into two basic categories: those which do not require a machining process to take place, and those which do. A parallel subdivision gives two more categories: those tests which merely indicate, for a given set of conditions, the relative machinability of two or more work–tool combinations (ranking tests), and those which indicate the relative merits of two or more work–tool combinations for a range of cutting conditions (absolute tests). The results of the former, whilst extremely useful in many circumstances, have two main disadvantages. Firstly, even if as hoped, a particular test indicates that material A machines better than material B which in turn machines better than material C, in most cases there is no indication of the magnitude of the differences because the measure of machinability does not, in general, correlate on a predictable scale with, for example, the life of the cutting tool under a given set of conditions. Secondly, even if the test does attempt to compare workpieces for a given set of cutting conditions, there is no guarantee that when the cutting conditions change the ranking will remain the same. The results of absolute tests are usually applicable for changes in cutting speed, certainly over the practical range, and in some cases also take account of changes in other cutting conditions and tool geometry. Thus, a machinability test can, in theory, be one of four types but, in practice, a non-machining test is always a ranking test whereas a machining test can be either a ranking or an absolute test.

Invariably, non-machining tests and machining ranking tests take less time to perform than absolute tests and are usually referred to as short tests. For virtually the whole of this century, long, absolute machinability tests have been carried out but until comparatively recently the conditions under which the tests were conducted—the tool and workpiece geometries, the cutting conditions, and the criteria of tool failure—were left to the discretion of the researcher and unfortunately this has resulted in vast amounts of machinability data each portion of which was pertinent at the time the test was carried out but which combined are difficult to correlate to form a cohesive body of data.

Since absolute tests give a more complete picture of the machining characteristics of a work–tool combination, they are, under many circumstances, considered to be superior to ranking tests and, as a consequence, considerable effort has gone into developing short absolute tests.

One relationship which must be established by anyone about to conduct machinability tests is that between the cost of the test and the potential benefit of carrying out the test. For example, one of the main uses of machinability tests is to evaluate incoming materials to a machine shop to ensure that the cutting tools to be used under specified conditions will not fail prematurely. However, before a test is considered to be suitable, it is important that the user establishes the likely incidence of premature tool failure if no screening of incoming material is carried out and, more importantly, how much this is likely to cost in lost production, scrapped tools and workpieces, etc. Whatever the cost, it is important that the cost of testing is less than the savings attributable to the testing and on the reasonable basis that more costly testing results in more effective workpiece evaluation, in general, an optimum test will be evident, i.e. there will be a level of testing for which the difference between the savings due to testing and the cost of testing is at a maximum. In practice there is a strong temptation either to do no testing or to do very thorough testing; this should be resisted.

Another important use of machinability testing is for the evaluation of new tool and workpiece materials. If, for example, a new workpiece material is being developed it is important to establish the machining characteristics of the material for a wide range of cutting conditions, tool materials and tool geometries and, perhaps more importantly, determine the consistency of the machining characteristics. To obtain information of the type required and to have the necessary confidence in the results would clearly require more extensive testing than that carried out for the previous examples but, again, care should be taken to ensure that, as far as possible, the cost of the testing is commensurate with the returns which are attributable to the results of the testing.

A specification for a standard long absolute machinability test has been available since 1977 (ISO 3685–1977) and is described in Chapter 6. The rest of this chapter will deal with short ranking and absolute machinability tests.

3.2. SHORT MACHINABILITY TESTS

A representative range of short-term tests is given in Table 3.1 which indicates the type of test, whether it is machining or non-machining, whether it is absolute or ranking, the type of operation it is suitable for, and the test parameter which is being investigated. These tests will now be described and an indication will be given of the type of results which are obtained and the relevance of the tests to practical assessments of machinability; the first five tests described are of the non-machining type whilst all the others require metal to be cut using a variety of cutting processes.

The first series of tests are all of the ranking type. The method of ranking workpiece materials relative to each other varies depending on the test being considered but usually the measure is one of three basic types:

(1) The cutting speed for which the tool life will be a particular time based on some reasonable criterion of tool life and for constant values of other cutting conditions and tool geometry. This ranking system is particularly relevant to turning and the two common measures are the V_{60} cutting speed for high-speed-steel cutting tools, and the V_{20} cutting speed for carbide cutting tools. These are, respectively, the cutting speed for a tool life of 60 min and the cutting speed for a tool life of 20 min and have been chosen because for many practical operations these would be of the same order as the economic tool life.

(2) The properties of a particular workpiece when compared with those of a reference material. In many of these types of test the reference material is considered to have a machinability 'index' of 100 and the test material is rated relative to this. Whilst an idea of magnitude can be gained from this type of test in that, for example, a material with a machinability index of 125 will be worse than one with an index of 200, it is not possible to relate the machinabilities in terms of tool life or cutting speed in proportion to the values obtained even for those cases where cutting tests have been carried

TABLE 3.1
SHORT-TERM MACHINABILITY TESTS

Test	*Type*				*Operation*			*Test parameter*		
	Machining	*Non-machining*	*Absolute*	*Ranking*	*Turning*	*Drilling*	*Sawing*	*Work-piece*	*Tool*	*Cutting fluid*
Chemical composition		*		*	*			*		
Microstructure		*		*	*			*		
Physical properties		*		*	*			*		
Rapid facing	*			*	*			*	*	*
Constant pressure	*			*	*	*	*	*		*
Taper turning	*		*		*			*	*	*
Variable-rate machining	*		*		*			*	*	*
Step turning	*		*		*			*	*	*
Degraded tool	*			*	*			*		
Accelerated wear	*			*	*			*		
Tapping	*			*						*
High-speed-steel tool wear rate	*		*		*			*		

out under specific relevant practical conditions. Usually the reference material is chosen to be of reasonable machinability and of the same basic type as the test materials but, importantly, it is often chosen for the consistency of its machining characteristics.

(3) The time taken for a predetermined event to take place. A good example of this would be the time to tool failure of a high-speed-steel tool when carrying out a facing test under controlled conditions. A further common example is the time for the surface finish of a machined component to deteriorate by a fixed proportion of the initial surface finish. As for the previous examples, the machinability is ranked against the values obtained for a reference material.

3.3. NON-MACHINING TESTS

3.3.1. Chemical Composition Tests

Many tests of this type have been developed where the results of other types of ranking test are correlated with the primary constituents of the materials under consideration. Most of the authors of this type of work acknowledge that the results obtained are only valid for materials of the same basic type and with the same thermal history but nevertheless these tests can, as can most of the ranking tests, be very useful for screening material prior to use provided the limitations of the test are understood. The two tests described here rank the materials relative to the V_{60} scale and a machinability index of 100, and are the result of work by Czaplicki (1) and Boulger *et al.* (2) respectively.

Czaplicki (1) determined the relationship between the V_{60} cutting speed and chemical composition as

$$V_{60} = 161{\cdot}5 - 141{\cdot}4 \times \%C - 42{\cdot}4 \times \%Si - 39{\cdot}2 \times \%Mn - 179{\cdot}4 \times \%P + 121{\cdot}4 \times \%S \qquad [3.1]$$

The author (1) specified this relationship for steels and claimed accuracies to within 8% of the values obtained from machining tests but clearly the range of materials considered and their thermal histories are restricted.

Boulger *et al.* (2) determined the relationship between their machinability index and chemical composition as

$$\text{Machinability index} = 146 - 400 \times \%C - 1500 \times \%Si + 200 \times \%S \qquad [3.2]$$

and this was for a range of free-machining steels. The significant influence of the level of residual elements on the V_{240} cutting speed when machining leaded free-machining steels has been described recently by Mills (3) and is shown in Fig. 3.1. This work shows that increasing the amounts of residual elements raises the strength of the ferrite which consequently increases the wear rate of the M2 high-speed-steel tools.

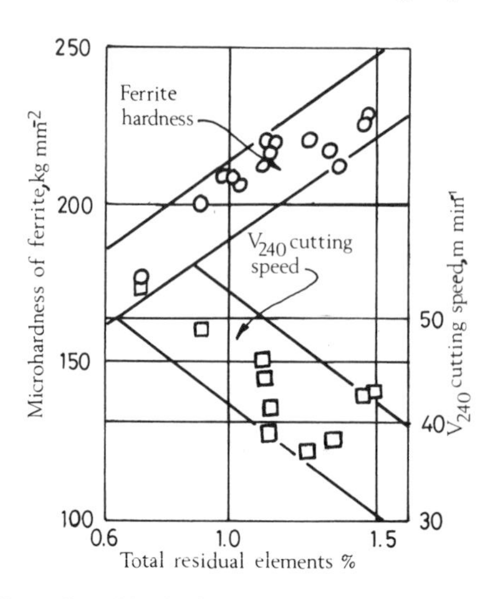

FIG. 3.1. Effect of total residual element level on ferrite hardness and V_{240} cutting speed (after Mills (3)).

3.3.2. Microstructure Tests

Some of the early work on the effect of the microstructure of low and medium carbon steels was carried out by Whittman (4), Woldman (5) and Robbins & Lawless (6). In general they concluded that uniformly distributed pearlite of large interlaminar spacing was the optimum microstructure for both turning and milling; similar conclusions were drawn by Boulger *et al.* (7) using their constant pressure test.

Perhaps the most useful evaluation of the effect of work material microstructure on the machinability of steels was carried out by Zlatin & Field (8). Their results have been summarised in Table 3.2 where it can be seen that steels which contain more than 50% pearlite combine good machining properties with high bulk hardness.

Field (9) and Murphy (10) both held the opinion that tool life is primarily dependent on the relative proportion of pearlite and ferrite and that an optimum ratio of the two exists.

TABLE 3.2
EFFECT OF MICROSTRUCTURE ON MACHINABILITY OF STEELS

Type of micro-structure	*Brinell hardness*	*V_{20} cutting speed ($m\ min^{-1}$) (carbide tool)*	*Machinability rating*	
			Relative life at constant speed (min)	*Relative speed at constant tool life ($m\ min^{-1}$)*
10% Pearlite + 90% Ferrite	100–120	290	8	22
20% Pearlite + 80% Ferrite	120–140	260	6	20
25% Pearlite	150	—	—	15
Spheroidised	160–180	180	5	14
50% Pearlite + 50% Ferrite	150–180	—	4	11
50% Fine pearlite + 50% Network ferrite	202	—	—	10
75% Pearlite + 25% Ferrite	170–190	140	3	11
100% Pearlite	180–220	145	2	11
Tempered martensite	240	—	—	8
,,	280–320	105	1	8
,,	350	—	—	6
,,	370–420	46	0·2	3

More recent work (11–13) indicates that for commercial materials the minor elements have a considerable effect on the machinability—perhaps the most widely studied of these is the effect of manganese and sulphur, in the form of manganese sulphide inclusions, in free-machining steels. Work by Chisholm & Richardson (11), Wilbur (12) and Lorenz & Evans (13) has shown that whilst the amounts of manganese and sulphur in a steel are clearly important, the shape, size and distribution of the manganese sulphide inclusions all affect the machinability. It has been suggested by Clark (14) that large uniformly distributed manganese sulphide inclusions give good machinability and that the addition of phosphorus up to a level of 0·7% helps to achieve this.

Whilst, in theory, an examination of microstructure should give a better indication of a material's machining properties than considering only chemical composition, two main problems exist for this type of test.

Firstly, it is difficult to measure, even approximately, the relative constituents of a material quickly and even when these can be determined the machinability ranking tends to be by subjective measures such as good, medium or bad. Secondly, the equipment and staff required to make the necessary measurements are not always available to companies which could make use of the information.

3.3.3. Physical Properties Tests

The search for a simple yet reliable criterion for assessing machinability based on physical properties led Henkin & Datsko (15) to develop a general machinability equation using dimensional analysis techniques. This was of the form

$$V_{60} \propto \frac{B}{LH_B}\left(1-\frac{Ar}{100}\right)^{1/2} \qquad [3.3]$$

where B=thermal conductivity of the material; L=a characteristic length; H_B=Brinell hardness of the material; and Ar=the percentage reduction in area of the material obtained from a conventional tensile test. The results of some of this work are presented graphically in Fig. 3.2 where it can be seen that for the steels chosen, a good correlation exists. Similar work by Janitzky (16) yielded the expression

$$V_{60} \propto \frac{D}{H_B Ar} \qquad [3.4]$$

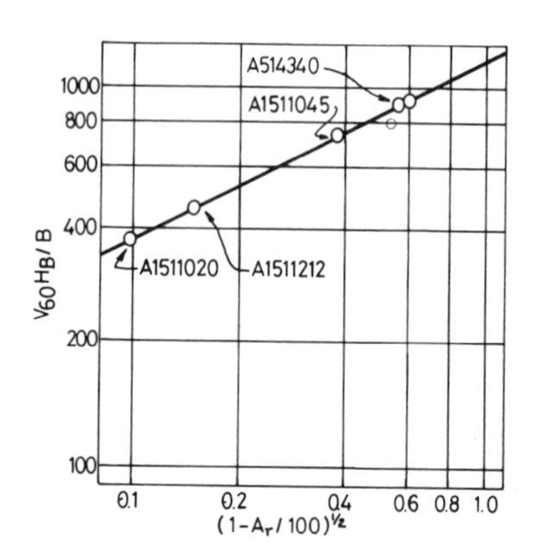

FIG. 3.2. Comparison between a general machinability equation and experimental data for a range of ferrous alloys (after Henkin & Datsko (15)).

where D is a constant dependent on the size of cut. Janitzky's results are shown graphically in Fig. 3.3 where, as for the results of Henkin & Datsko (15), a good correlation exists. Again, provided the necessary equipment required to carry out the tests is available and provided the composition of the materials under study is similar, there is a very good chance that the results obtained for the V_{60} cutting speed can be determined to an acceptable accuracy.

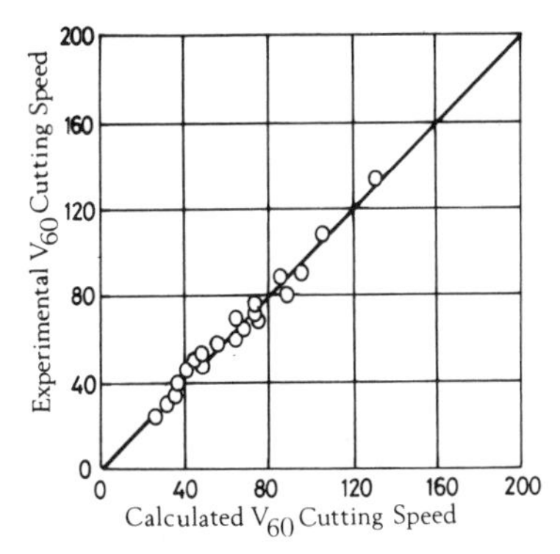

FIG. 3.3. Comparison between experimental tool lives and calculations based on mechanical property data for twenty steels (after Janitzky (16)).

3.4. MACHINING TESTS

3.4.1. The Constant Pressure Test

Perhaps the best known of all the machining ranking tests is the constant pressure test developed by Boulger *et al.* (7). In the turning test the feed force is kept constant by using a specially adapted lathe and the rate of feed achieved by cutting the material under test using a tool of pre-determined geometry is taken as a measure of the machinability of the material. It can be shown that the technique basically measures chip–tool friction but since this is closely related to the temperatures in cutting which, in turn, have a pronounced effect on the rate of wear of the cutting tool, it could be expected that the method will yield useful results. In practice, it has been found that the method is particularly useful for evaluating the relative merits of free-machining steels.

The constant pressure test has also been adapted to include drilling and sawing tests. As could be expected, the limitations which apply to the turning test also apply to these. The main disadvantage of the test, in practice, is that, to obtain results quickly, a specially adapted machine tool is required on which to carry out the tests. However, in, for example, turning tests a standard lathe and a two-component metal cutting dynamometer could be used in which case it would be necessary to plot the feed force versus feed relationship to determine the feed for a specific feed force; this testing method would, necessarily, take longer than that previously mentioned. Typical results of a series of constant pressure tests are shown in Fig. 3.4.

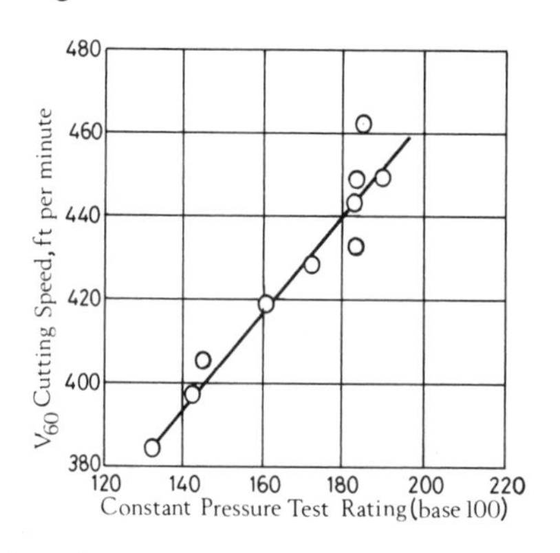

FIG. 3.4. Comparison between machinability ratings for two types of test.

3.4.2. The Rapid Facing Test

A ranking test which is said to give a useful measure of the machinability of a material is the facing test as proposed by Kraus & Weddell (17) and Lorenz (18). In this turning test the workpiece material is faced from the centre, preferably on a large diameter bar and the measure of machinability is the distance the tool travels radially outwards or the cut length up to the point of tool failure compared with the same test carried out on a reference material. The test is limited to high-speed-steel tools since tool catastrophic failure is not a valid criterion for carbides. Whilst the test is quick and can be carried out on a standard lathe, the two biggest limitations are that, firstly, the diameter of the material to be used in practice may be considerably less than would be appropriate for carrying out the test and, secondly, unless the material is homogeneous, the test will assess the properties of the material at a range of diameters and this would not necessarily be indicative of the properties of the material at the diameter at which it is to be cut. The latter disadvantage can be at least partially overcome by drilling out the centre of the workpiece to a suitable diameter.

3.4.3. Tapping Tests

It was thought appropriate to mention here a test which is not concerned with assessing the quality of either the tool or the workpiece material but with assessing cutting fluids. For many years, a standard test used by cutting tool manufacturers, whilst not strictly a short test, has been the tapping test where the merits of the fluid are judged by either the number of holes tapped to a suitable size or the amount of flank wear generated on the tap per hole tapped. The test is simple and does not consume large quantities of material but if the amount of time required to carry out the test is to be kept reasonable some form of automatic operation, such as the use of a numerically controlled machine tool, would be helpful. It is said by the cutting fluid manufacturers that the results of tapping tests correlate very well with the results obtained in practice for a variety of machining operations.

3.4.4. Degraded Tool Tests

Many tests have been developed which attempt to assess the machinability of workpieces by cutting them with softened cutting tools. In a piece of work by Mills & Redford (19) a range of leaded α/β brasses were assessed using a softened high-speed tool and it was found that the hardness of the tool was extremely critical in that a tool slightly too soft would fail catastrophically in a very short period of time whereas a tool which was slightly too hard would not wear at all. Undoubtedly, when the test material is more difficult to machine and when the tool does not have to be drastically softened, the hardness of the tool would be less critical.

For this type of test the normal criterion of tool failure is the same as for most tool wear type tests, i.e. either a given amount of tool flank or tool crater wear, or catastrophic failure of the tool. If a range of materials are to be compared then the basic assumption of the test is that the machinability of a range of materials will remain in the same order if full-strength tools were to be used. Unfortunately, this is not necessarily true, but, as for most short ranking tests, if care is exercised and the tests are limited to similar materials reasonable results should be obtained. If the test is to be used for assessing a particular material relative to some standard then the relationship between the results of tests for the softened tool and for a standard tool would have to be determined. This type of test should give good results but in practice difficulty is usually found in maintaining the consistency of the degraded tool when the tool has to be changed.

3.4.5. Accelerated Wear Tests

An alternative to softening the cutting tool as a means of achieving higher wear rates is to uprate the cutting conditions, in particular cutting speed. Since for high-speed-steel tools the tool life is usually proportional to between the fourth and fifth power of cutting speed and for carbides to between the second and third power of cutting speed, significant increases in wear rate can usually be achieved for modest increases in cutting speed. Unfortunately, whilst the Taylor relationship ($VT^{-1/k}=C$) is usually valid in the practical cutting speed range, there are often deviations from this when cutting at very low or very high speeds. For high speeds this is often attributable to a change in state of the material being cut because of the very high cutting temperatures involved. Because of this effect, it is often unwise to extrapolate tool life cutting speed data particularly if the intention is to obtain quantitative information about the life of the cutting tool for cutting speeds outside the range of the tests. Thus, although the accelerated wear test is basically an absolute test it is more likely to give useful information about the ranking of a range of similar materials or, alternatively, could be very useful as a method of assessing the machinability of a material against a predetermined standard.

3.4.6. High-Speed-Steel Tool Wear Rate Test

A recent study (20) has been carried out on the machinability of free-machining steels using absolute tests and where the test parameters were the quantities of minor elements included in the steels. The tests were carried out using high-speed-steel tools and because accurate information was required the long term test as specified by the International Standards Organisation (ISO 3685-1977) was used. As was expected, the time to catastrophic failure decreased as the cutting speed was increased but, more significantly, the amount of wear on the tool at the onset of catastrophic failure decreased as the cutting speed was increased. Thus, extrapolation of tool wear–cutting time data which could be done for a limited flank wear tool life criterion was not possible. However, it was found that when the flank wear just prior to failure versus tool life relationship was plotted on a linear–log scale the relationship gave a straight line (Fig. 3.5) and for the materials tested yielded an equation relating tool life to wear rate in the secondary wear region together with a constant which was the flank wear at the end of the primary wear region. This and the rate of wear in the secondary wear region were determined for five materials after a period of 60 min testing and the results obtained for tool life

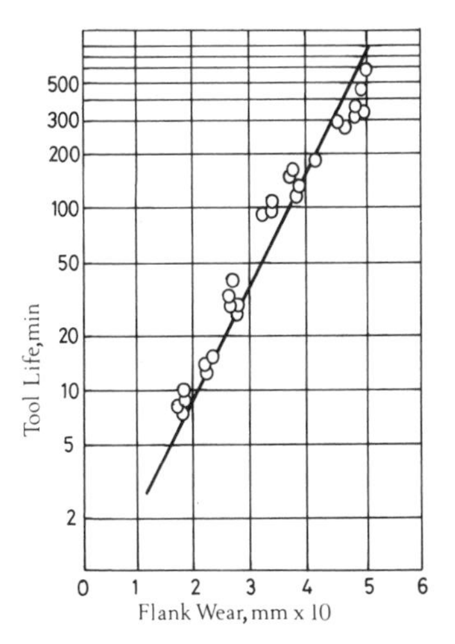

FIG. 3.5. Relationship between flank wear at the end of the secondary wear stage and tool life for an M2 high-speed steel (after Akhtar *et al.* (20)).

were compared with those obtained from completing the full test; the agreement was very good and typically the testing time and material consumed were reduced to 20% of that required for a full test. Further, although the workpieces used were of the same family they were sufficiently different for it to be reasonable to suggest that the relationship obtained was a characteristic of the tool rather than the workpiece and that it should therefore be possible to use the test for a wider range of workpieces.

3.4.7. Taper Turning Test

This test is one of three basically similar tests, the others being the variable-rate test and the step turning test. The test procedure was proposed by Heginbotham & Pandey (21) and essentially, as the name implies, a tapered workpiece is machined along the taper in order to produce cutting conditions where the cutting speed increases in proportion to the time of cut. It can be shown that, with a knowledge of the initial cutting speed, the rate of change of cutting speed with time, the time of cut, and the amount of wear on the tool flank, if two tests are carried out, expressions can be formulated which will give the values of the constants k and C_1 in the equation $VT^{-1/k} = C_1$. It is suggested that the principal advantages of the test are that it is relatively short whilst at the same time it is comprehensive in that every cutting speed in the range chosen is used to produce wear on the tool. Results presented by the authors (21) showed excellent agreement with those obtained by conducting a long machinability test. An independent assessment of the test (22) indicated that consistent results could be obtained but unfortunately these did not agree with the values obtained for a long test due to the changing angle of the flank throughout the test. Other disadvantages of the test are that tapered specimens have to be prepared, a taper turning attachment is needed on the machine tool, and, because the tool is cutting at different diameters, the results may not be representative of how the material would behave if all the cutting were carried out at one diameter.

Despite its disadvantages, the test introduced a new concept into short-term absolute machinability testing and led to the variable-rate test.

3.4.8. Variable-Rate Machining Test

The principal difference between this test and the taper turning test is that the increase in cutting speed with time is achieved by axially turning a workpiece and gradually increasing the rotational speed of the machine. As was the case for the taper turning test, the constants k and C_1 of the tool life equation can be determined by conducting two variable-rate tests. The analysis relating the constants k and C_1 to the machining conditions is presented in Appendix 2.

The advantages of the variable-rate test over the taper turning test are that a standard workpiece can be used and the problems of machining at different diameters are not present. The disadvantage of the test is that some means of progressively increasing the rotational speed of the machine as the cut takes place is required—at the time the initial testing was carried out, this was achieved by an electromechanical system which would not be very practical in an industrial situation. Now, it would be easy to do this type of test on a numerically controlled (NC) lathe but this type of machine may not always be available to the tester.

3.4.9. Step Turning Test

The step turning test, as proposed by Kiang & Barrow (23), was developed to overcome the practical difficulties of both the taper turning and the variable-rate tests by formulating the tests so that they can be carried out on a standard lathe. This is achieved by using a range of discreet speeds. The results obtained by the authors (23) showed good agreement with those obtained from a long test and the analysis relating machining conditions to the constants k and C_1 is given in Appendix 2.

Comparisons between experimental and theoretical results for taper turning, variable-rate and step turning tests are shown in Figs 3.6, 3.7 and 3.8. The variable-rate and the step turning tests have both given indications that good data can be obtained without incurring the costs of a long test but these tests were carried out under laboratory conditions with good control over the consistency of the tool geometry and material, the workpiece material, and the cutting conditions. It is felt that in a practical situation where these parameters may be more difficult to

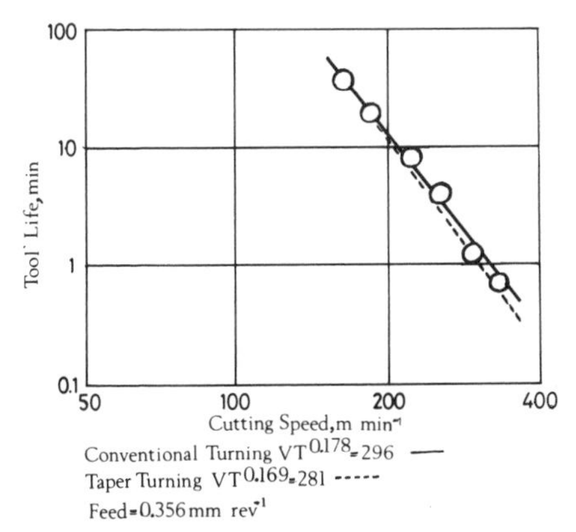

FIG. 3.6. Comparison between the cutting speed–tool life relationship for conventional and taper turning (after Heginbotham & Pandey (21)).

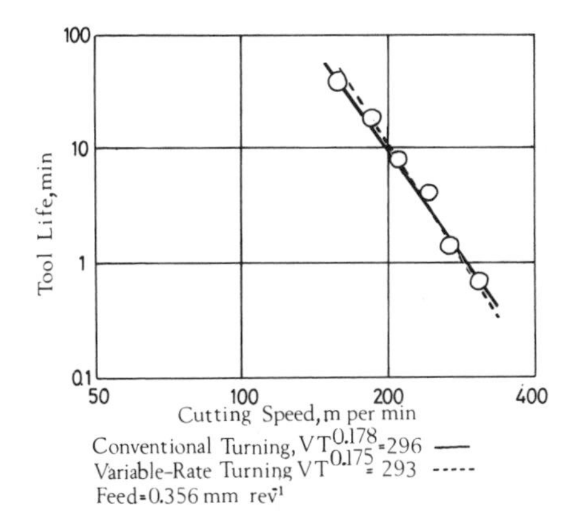

FIG. 3.7. Comparison between the cutting speed–tool life relationship for conventional and variable-rate turning (after Heginbotham & Pandey (21)).

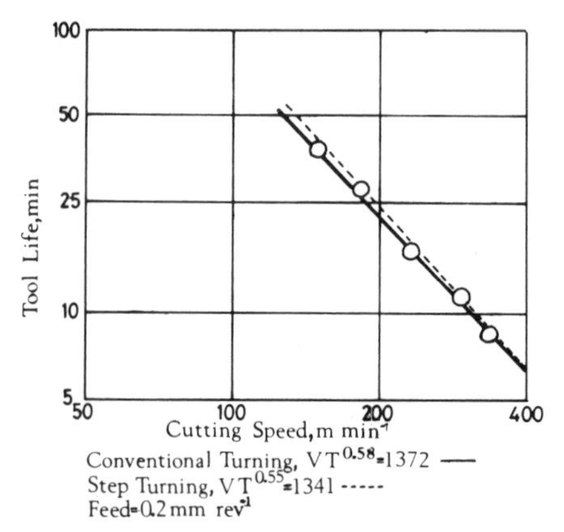

FIG. 3.8. Comparison between the cutting speed–tool life relationship for conventional and step speed turning (after Kiang & Barrow (23)).

control, the confidence which can be placed in the values obtained is a function of the amount of data which is collected and that, in the limit, there is no substitute for the long test. However, since the objective when carrying out most absolute tests is often concerned with tool and workpiece development and under these circumstances good control should be possible, there is a very good argument for using the short tests.

3.5. COMBINATION OF MACHINING PARAMETERS

It is well known that by far the most significant machining parameter when considering tool life is the cutting speed. However, feed, depth of cut, clearance angle, tool nose radius etc., all have an effect on tool life and if these were all to be considered the number of tests required to even adequately cover a reasonable range of cutting conditions would be prohibitive even if short tests were to be used. As a consequence, significant effort has gone into trying to combine the machining parameters in a meaningful way such that the results from a few tests may be applied to a wider set of cutting conditions.

Introducing the chip equivalent concept, Woxen (24,25) showed experimentally as well as theoretically that the chip equivalent is a basic quantity which together with cutting speed determines the life of a cutting tool. The chip equivalent is defined as the ratio between the engaged cutting edge length and the chip cross-sectional area. There are many formulae for calculating the chip equivalent depending on the cutting conditions. Whitehead (26) has developed expressions for round, square and triangular throw-away tools for all the usual combinations of feed and depth of cut.

There is considerable evidence to suggest that, for a given workpiece–tool combination, the tool life is a function of the cutting temperature regardless of the cutting conditions. Cutting temperature can be measured in a variety of ways but the most common method is to use a work–tool thermocouple which gives a reading which approximates to a mean cutting edge temperature. A typical tool life–temperature relationship is shown in Fig. 3.9 where it can be seen that for variations in two cutting parameters, speed and feed, a unique temperature–tool life relationship exists of the form $\theta T^{n}=C$, where θ is the work–tool thermocouple reading, T is the tool life, and n and C are constants. Typically, for dry turning of steel with a high-speed-steel tool, n is in the range 0·05 to 0·1, i.e. a small variation in cutting temperature has a marked effect on tool life.

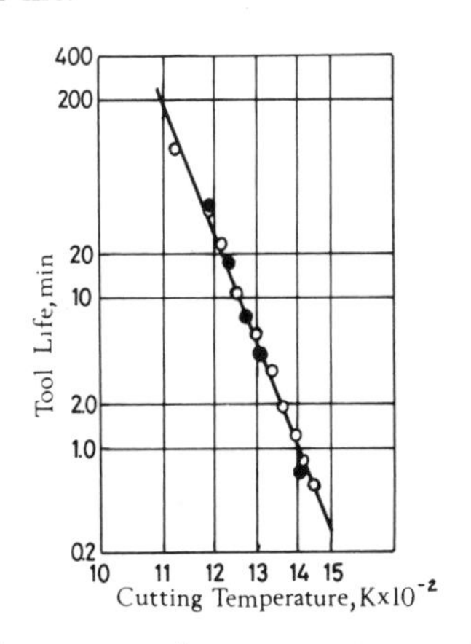

FIG. 3.9. Relationship between cutting temperature and tool life for a carbide tool machining a heat-resistant alloy (after Takeyama & Murata (27)).

If valid, tool temperature measurement offers not only a rational approach to tool life assessment but also one that can save a great deal of time when evaluating the machinability of materials. Once the relationship between temperature and tool life has been ascertained for a given work–tool combination, the effect of changing cutting conditions can quickly be determined. As with many other methods of machinability assessment, specialist equipment is required although, in its simplest form, the work–tool thermocouple can be fitted to a standard lathe without too much difficulty.

3.6. MACHINABILITY ASSESSMENT FOR PROCESSES OTHER THAN SINGLE POINT TURNING

3.6.1. Introduction

In keeping with its importance, the major portion of this chapter has been devoted to the assessment of machinability in turning since, for a variety of reasons, this has been the main process for which machinability data has been generated. Further, the process lends itself to test work in that it is generally a continuous-cutting operation, the range of practical tool geometries, particularly for roughing operations, is limited and simple, chip flow patterns are predictable, and the tool itself is relatively cheap. The latter is particularly significant in that it results in optimum metal removal rates being high, and significant economies resulting from operating at or near optimum conditions. Conversely, of the other major metal removal processes, milling and planing/shaping, planing is basically a low speed process which is not particularly affected by the generation of machinability data and for which turning data will generally be valid, whilst milling is a complex process with many types of cutter, many types of operation, and until relatively recently with the advent of milling cutters with indexable inserts, very expensive tools. The latter is a most significant factor in that metal removal rates have not been improved by optimising cutting speed since for expensive high-speed-steel tools premature tool failure can prove to be very costly. Thus, conservative cutting speeds which result in long tool lives tend to be used and these have, relatively, little effect on the metal removal rate. However, much work has been conducted into the machinability of tool–workpiece combinations for processes other than turning and the two most common, drilling and milling, will be discussed briefly.

3.6.2. Machinability Assessment in Drilling

Both ranking and absolute drilling tests have been conducted in the past and the common ranking test has been the constant pressure test where a given thrust is applied to a drill and the measure of machinability is the rate of penetration of the drill relative to the rate of penetration when drilling a reference material. As is the case for constant pressure lathe tests, the test is particularly suited to ranking free-machining steels. Another common ranking test for drilling is based on the number of holes which can be drilled before the hole produced does not conform to a specified tolerance. Again the measure of machinability is usually relative to the number of holes drilled in a reference material.

Conventional absolute machinability tests have also been carried out for drilling and the tool failure criterion has been either catastrophic failure or some measure of drill tip wear. Unfortunately, in drilling, and particularly when using small drills, there is a high probability of premature failure which is caused not by cutting edge failure but by drill breakage. It has been shown by several authors that perhaps the most significant drill parameter which affects the life of the drill is its length—testing has indicated that changing the length of the drill has a marked effect on the incidence of drill breakage.

On the assumption that drill breakage is reduced by limiting the length of the drill considerable work has been done on the evaluation of machinability in drilling by several authors using short drills.

The common element in all the work examined, whatever the criterion used, was that drill life testing is inconsistent and this is attributed, primarily, to the marked effect on drill life of very small changes in drill geometry. Williams & McGilchrist (28) discounted increases in drilling torque or quality of hole produced as reasonable criteria of tool life and used a catastrophic failure criterion. Lorenz (29) in extensive testing indicated that the appropriate British Standard for an acceptable drill was inadequate in that virtually any drill, almost regardless of quality, could meet the specification. He suggested that a technique for testing based loosely on response surface methodology, first proposed by Wu (30) and later used by Williams & McGilchrist (28), should be used; the criterion of tool failure was catastrophic.

Other researchers (31, 32, 33) chose to measure corner wear and margin wear and assumed a limited value of these as an acceptable measure of tool life; as mentioned above, acceptable repeatability even under controlled laboratory conditions was difficult to achieve.

In conclusion, it can be said that where premature drill failure would prove to be costly, and because drill life is very variable, very conservative cutting conditions should be used and drill replacement carried out far more regularly than would be indicated as necessary from 'average' figures for tool life.

3.6.3. Machinability Assessment in Milling

It is indicative of the difficulty of establishing the many standard tests which would be necessary to adequately deal with the whole range of milling operations using various types of milling cutter that only now is the International Standards Organisation (ISO) looking at one example of milling and great difficulty is being encountered in establishing a standard meaningful test for the type of cutter under investigation. Various types of milling test have been carried out and, since absolute machinability tests tend to be costly, the emphasis has been placed on developing methods of testing which take little time.

Yellowley (34) developed a technique which allows the tool life for given conditions to be established for a variety of operations, including milling, from a knowledge of the stress-modified temperature at the start of cutting. In a further paper, Yellowley & Barrow (35), when investigating specifically the milling process, introduced the concept of a thermal fatigue parameter and showed that this was of great importance when attempting to establish the tool life in milling.

Colding (36) extended the chip equivalent concept, developed by Woxen (25) for turning, to include milling; tests for plane and face milling operations yielded encouraging results.

It is likely that in the future with the increasing use of milling cutters with indexable inserts, more effort will be put into generating machinability data for milling since a survey has indicated that typical metal removal rates in milling are only a fraction of those which should be used for either minimum cost or maximum production rates.

3.7. MACHINABILITY ASSESSMENT RELATING ONE PROCESS TO ANOTHER

As has already been indicated, machinability determination is very much process- and cutting condition-dependent and it is difficult to relate results from one set of conditions to what may happen under different conditions using the same process. It is not surprising, therefore, that to try to relate machinability data obtained from one process to another is virtually impossible unless the processes involved are effectively the same. Thus, for example, turning data can be related to shaping, planing, and boring because they are all single point, effectively continuous-cutting, operations. However, to try to relate turning data to milling (discontinuous process with complex tool geometry) or drilling (complex tool geometry) is extremely difficult and the confidence which could be placed in the observations would be low. Although some attempts have been made to achieve these objectives, the results obtained to date have been generally disappointing.

3.8. ON-LINE ASSESSMENT OF TOOL WEAR

In most tool wear testing methods a significant part of the total testing time is devoted to the measurement of the wear whatever the type of wear being investigated and, clearly, in practical cutting operations this cannot be tolerated. If, in practice, the tool wear at any instant could be measured quickly, without interrupting the normal production sequence of the machine tool, the large safety factors which are applied to the useful life of the cutting tool to prevent premature tool failure either by reducing the time of use or, more usually, reducing the cutting speed, would be unnecessary. Under these circumstances tools would in general cut for longer and the cost of the tool and its replacement per unit volume of metal removed would be reduced. Further, if the tool wear could be measured regularly without penalty then the tool wear rate could be assessed at any time and from a knowledge of this, together with the machine characteristics and the various cost parameters, it would be possible to estimate the optimum cutting conditions for a chosen criterion such as minimum cost or maximum production rate.

Optimisation of the cutting conditions to meet a specific enonomic criterion is usually referred to as 'adaptive control' and this has been the subject of much research effort in recent years. Limited capability adaptive control machine tools are now commercially available and these in common with more ambitious systems rely in some way on at least limited knowledge of the tool wear rate at any instant together with the amount of wear.

Many methods of measuring tool wear under production conditions have been investigated and some of these will now be discussed. The methods developed fall into one of two broad categories: those methods where an attempt is made to physically measure the wear, and those which rely on an indirect method of wear measurement. The former are invariably restricted to measurement of flank wear whilst it is claimed that some of the latter indirect methods are capable of differentiating between flank and crater wear. The former methods can clearly only be used if the tool is out of cut at fairly regular intervals whereas the latter can only be used in the more practical case of the tool being in cut.

Typical operations for which direct measurement methods have been used are short machining time work such as occurs on capstan lathes, automatic lathes and numerically controlled lathes where the tool to be measured can be stationed in some convenient position whilst measurement takes place. The method of measurement can make use of an optical microscope or more commonly a video camera which has the capability of recording the picture for more leisurely evaluation. The three main disadvantages of this method of assessment are:

(1) Some potentially useful machining time is lost whilst the measurement is being carried out although, with care, this can be reduced to an insignificant proportion of the total machining time.
(2) The recording equipment is expensive and generally an operator is required to 'decode' the information. If automatic decoding is to be used using pattern recognition techniques then the process becomes even more expensive.
(3) In some instances cutting debris will tend to obscure the view of the wear land and unless thorough cleaning techniques are employed, some of the results obtained will be misleading.

Despite these disadvantages, the method has found limited use in practice and under some circumstances it can be shown to be economically justifiable.

Indirect methods of measuring tool wear have been developed along several lines but the most common have been concerned with the measurement of changes in power consumed, workpiece geometry, tool forces or tool junction temperature.

Direct power consumption methods have had little success because of the relatively small changes in total power consumed as the tool wears.

A technique which has proved successful in metal forming operations is that of comparing a 'standard' power–time curve with that being generated—however, when this is applied to metal cutting it is difficult to differentiate between the small fluctuations due to wear and those due to other influences such as variations in the properties of the workpiece material.

Several techniques have been developed which attempt to determine changes in workpiece geometry due to tool wear but even under ideal conditions they are only suitable for measuring uniform flank wear. Various direct and indirect distance measuring systems have been tried but the former tend to be too fragile for production conditions whilst the latter, even if capable of withstanding production conditions, invariably fail because under most circumstances the transient workpiece surface which is being measured has irregularities which are of the same order of size as the change in geometry which is being measured.

Another method which has been tried is that of using a secondary tool set to cut when a given amount of flank wear has been generated. This has also generally proved to be unsuccessful because it is difficult to set the primary and secondary tools relative to each other and, again, as for the method described above, the changes in geometry to be measured are small and surface irregularities on the transient workpiece surface lead to a false impression of the state of the tool.

Measurement of tool forces to detect both flank and crater wear have been studied extensively by Micheletti *et al.* (37), De Filippi & Ippolito (38) and Konig *et al.* (39). The problem facing these investigators has been that of separating the effects of flank and crater wear. It is well known that a flank wear land tends to increase both the tangential cutting force and the feed force whilst a crater tends to produce a decrease in these forces.

Micheletti *et al.* (37) suggested that for flank wear a measure of the shear force acting on the flank of the tool would be highly significant. Unfortunately, the isolation and measurement of this force is very difficult and would not be practicable. De Filippi & Ippolito (38) tried to relate increases in feed force with machinability using a short test. As could be expected, significant wear is necessary before the increase in feed force can be detected with reasonable accuracy and under many circumstances the flank wear would be too large. Further, the results obtained by De Filippi & Ippolito (38) for flank wear and the corresponding feed force exhibited considerable scatter and if these were to be used to assess flank wear rate meaningless data would be obtained. Similar work carried out by Konig *et al.* (39), specifically aimed at developing a short machinability test but which was based on measurements of feed force, appeared to be more successful. However, it is unlikely that the sensitivity required for adaptive control would be evident.

Even if force measurement techniques were successful, one of the main problems of attempting to relate force to tool wear is that changes in cutting speed, feed, depth of cut, and tool geometry all affect the cutting and feed forces and as a consequence any assessment would be relative to a datum which would vary depending on the operation. Conversely, as has been previously mentioned, there is much evidence to suggest that the relationship between cutting temperature and tool life is unique for a given tool–workpiece combination and tool geometry and it would be reasonable to suppose that increases in cutting temperature due to increases in flank wear would also be unique and independent of cutting conditions. Unfortunately, work carried out to date does not appear to substantiate this observation and whilst there is no doubt that the cutting temperature, as measured by a work–tool thermocouple, does vary with the amount of flank wear, the results obtained have been erratic and unsuitable for accurate measurements of tool wear rates under production conditions.

In conclusion it may be said that a prerequisite of effective adaptive control is a method of accurately estimating tool wear whilst the tool is cutting and that, to date, no such method exists.

REFERENCES

1. CZAPLICKI, L. (1962). L'usinabilite et al coupe des metaux, *Res. Soc. Roy. Belge Ingeniere*, **12**, 708–36.
2. BOULGER, F. W., MOORHEAD, H. A. & GAVEY, T. M. (1951). Superior machinability of MX steel explained, *Iron Age*, **167**, 90–5.
3. MILLS, B. (1980). Effect of residual elements on the machinability of leaded free machining steels, *Phil. Trans. R. Soc. Lond.*, **A 295**, 87–8.
4. WHITTMAN, G. P. (1945). Studies of the machinability of carbon and alloy steels, *Mech. Engrs.*, **67**, 575–83.
5. WOLDMAN, N. E. (1947). Good and bad structures in machining steels, *Materials and Methods*, **25**, 80–6.
6. ROBBINS, F. J. & LAWLESS, J. J. (1955). Heat treating—for better machinability, *Iron Age*, **176**, 94–7.
7. BOULGER, F. W., SHAW, M. L. & JOHNSON, H. E. (1949). Constant pressure lathe test for measuring machinability of free machining steels, *Trans. A.S.M.E.*, **71**, 431–8.
8. ZLATIN, N. & FIELD, M. (1950). Evaluation of machinability of rolled steels, forgings and cast irons, In: *Machining theory and practice*, A.S.M., Cleveland, Ohio.
9. FIELD, M. (1963). Relationship of microstructure to the machinability of wrought steels and cast iron, *Proc. Int. Prod. Eng. Res. Conf.*, Paper 20, Carnegie Inst. of Tech., Pittsburgh, Pennsylvania, Sept. 1963.
10. MURPHY, D. W. & AYLWARD, P. T. (1965). Measurement of machining performance in steels, *Metalworking of steels*, AIME, 49–82.
11. CHISHOLM, A. W. J. & RICHARDSON, B. D. (1965). A study of the effect of non-metallic inclusions on the machinability of two ferrous materials, *Proc. Inst. Mech. Engrs.*, Feb. 1965.

12. WILBUR, W. J. (1970). *The effect of manganese sulphide inclusions on the mechanics of cutting and tool wear in the machining of low carbon steels*, PhD Thesis, University of Salford, UK.
13. LORENZ, G. & EVANS, P. T. (1967). Improving the machinability of low alloy steels by better metallurgical control, *A.S.T.M.E. Inst. Machining and Tooling Symp.*, Sydney, 1967.
14. CLARK, W. C. (1964). Which free machining stainless steels?, *Metalworking Production*, **180**, 43–7 and 68–71.
15. HENKIN, A. & DATSKO, J. (1963). The influence of physical properties on machinability, *Trans. A.S.M.E., Journal of Engineering for Industry*, November, 321–7.
16. JANITZKY, E. J. (1944). Machinability of plain carbon, alloy and austenitic steels and its relation to yield stress ratios when tensile strengths are similar, *Trans. A.S.M.E.*, **66**, 649–52.
17. KRAUS, C. E. & WEDDELL, R. R. (1937). Determining the tool-life cutting speed relationship by facing cuts, *A.S.M.E., Fall Meeting*, Erie, Pennsylvania, Oct. 4–6, 1937.
18. LORENZ, G. (1970). *Determination of comprehensive machinability equations by means of rapid facing tests*, S.M.E., MR 70–177.
19. MILLS, B. & REDFORD, A. H. (1978). A machining study of α/β leaded brasses, *Annals of C.I.R.P.*, **27**, 45–8.
20. AKHTAR, S., REDFORD, A. H. & MILLS, B. (1976). A short-time method for the assessment of the machinability of low carbon free machining steels, *Int. J. Mach. Tool Des. Res.*, **16**, 71–5.
21. HEGINBOTHAM, W. B. & PANDEY, P. C. (1966). Taper turning tests produce reliable tool life equations, *Proc. 7th M.T.D.R. Conf.*, 515.
22. ATKINSON, L. (1973). *An assessment of the taper turning test*, Internal Report, University of Salford, UK.
23. KIANG, R. S. & BARROW, G. (1971). Determination of tool life equation by step turning test, *Proc. 12th M.T.D.R. Conf.*, 379–86.
24. WOXEN, R. (1934). New theory for cutting metals with shearing tools, *Teknisk Tidskrift*, **61** (16), 43–4.
25. WOXEN, R. (1932). Theory and equations for the life of lathe tools, *Ingeniors Vetenskaps Akademien Handlingar*, **119**, 73.
26. WHITEHEAD, J. (1970). *The chip equivalent method of cutting tool life equations*, MSc Dissertation, University of Manchester, UK.
27. TAKEYAMA, H. & MURATA, R. (1963). Basic investigation of tool wear, *J. Eng. for Ind.*, 33–8.
28. WILLIAMS, R. A. & MCGILCHRIST, C. A. (1972). An experimental study of drill life, *Int. J. Prod. Res.*, **10**, 175–91.
29. LORENZ, G. (1977). A contribution to the standardisation of drill performance tests, *Annals of C.I.R.P.*, **25**, 39–43.
30. WU, S. M. (1964). Tool life testing by response surface methodology, *Trans. A.S.M.E., Journal of Engineering for Industry*, **86**(2), 105–16.
31. LENZ, E., MAYER, J. E. & LEE, D. G. (1978). Investigation in drilling, *Annals of C.I.R.P.*, **27**, 49–53.
32. KANAI, M. & KANDA, M. (1978). Statistical characteristics of drill wear and

drill life for the standardisation performance tests, *Annals of C.I.R.P.*, **27**, 61–6.

33. KALDOR, S. & LENZ, E. (1980). Investigation of tool life of twist drills, *Annals of C.I.R.P.*, **29**, 23–7.
34. YELLOWLEY, I. (1975). *The assessment of machinability*, S.M.E. Technical Paper MR75-147.
35. YELLOWLEY, I. & BARROW, G. (1975). The influence of thermal cycling on tool life in peripheral milling, *Int. J. Mach. Tool Des. Res.*, **16**, 1–12.
36. COLDING, B. N. (1961). Machinability of metals and machining costs, *Int. J. Mach. Tool Des. Res.*, **1**, 220–48.
37. MICHELETTI, C. F., DE FILIPPI, A. & IPPOLITO, R. (1968). Tool wear and cutting forces in steel turning, *Annals of C.I.R.P.*, **16**, 353–60.
38. DE FILIPPI, A. & IPPOLITO, R. (1972). Analysis of the correlation among cutting force variations, chip formation parameters and machinability, *Annals of C.I.R.P.*, **21**, 29–30.
39. KONIG, W., LANGHAMMER, K. & SCHEMMEL, H. V. (1972). Correlations between cutting force components and tool wear, *Annals of C.I.R.P.*, **21**, 19–20.

Chapter 4

TOOL MATERIALS

4.1. HISTORICAL BACKGROUND

The earliest uses of materials including steels were for the manufacture of rudimentary tools and it is interesting, therefore, to consider the chronological developments of tool materials. Naturally occurring ceramics were amongst the earliest materials used for tools although it is only in recent years that controlled manufacturing methods have enabled ceramic tools of good quality to be used for metal machining. Some of the major developments in the evolution of tool materials are listed in Table 4.1.

The Wootz steels of India were made from sponge iron in a crude forge. After hammering iron and remelting with wood a uniform steel was obtained. It is almost certain that steelmaking was practised by the Chinese as long ago as 1000 BC. Later the famed Damascus steel was made by welding alternate thin strips of steel and soft iron by twisting and working; the Toledo steels of Spain followed slightly later. During the so-called Dark Ages steel was manufactured by heating iron in contact with charcoal. In Europe prior to 1740, steel was made by carburising; after carburising, strips were broken into short lengths and stacked and reforged at red-heat. Although the product of this process was an improvement over previous steel it still lacked the uniformity of structure to allow through hardening. In 1740 Benjamin Huntsman rediscovered the crucible method of steel melting which produced a uniform product which responded consistently to heat-treatment. The differences in the forms of iron and steel due to differences in carbon content were not established until 1820 by Karsten. The next major

TABLE 4.1
CHRONOLOGY OF MATERIALS USED FOR CUTTING METALS (1)

Year	*Development*
BC 350	Wootz steels from India
AD 300	Damascus and Toledo Steels
Dark Ages (5th–8th centuries)	Cementation process (carburised iron)
1740	Rediscovery of melting by Huntsman
1868	Mushet's special steel (2%C, 7%W, 2·5%Mn)
1898	Taylor and White's work on high-speed steel
1903	Prototype of modern high-speed steel
1910	First 18%W–4%Cr–1%C composition
1915	Cast cobalt alloys
1923	Cemented carbides
1939	High C, high V, super high-speed steels
1940–1952	Substitution of Mo for W in high-speed steel
1953	Resulphurised high-speed steels
1957	Ceramic tools based on alumina

discovery in the development of tool steels was Mushet steel developed in 1868. Mushet found that a steel containing tungsten and chromium became hard without quenching. This steel found use as a cutting tool primarily in turning and planing operations since it had, compared with other steels of the time, excellent wear resistance and consequently required only infrequent regrinding. Experiments by Taylor and White of the Bethlehem Iron Company (Bethlehem, Pennsylvania, USA) on self-hardening alloy steels led directly to the development of superior heat-treatments. They found that high temperature heat-treatments (just below the melting point) led to superior cutting properties.

In the early 1900's, Taylor and White were recommending a steel containing 0·7%C and 14%W which was the forerunner of the present tungsten-based high-speed steels. At this time vanadium was also introduced as an alloying element in high-speed steels. By 1920 the three main grades of high-speed steel available were: 18%W–4%Cr–1%V; 14%W–4%Cr–2%V; and 18%W–4%Cr–1%V plus cobalt. The wide variety of high-speed-steel grades available today includes the molybdenum grades which were developed because of the shortage of tungsten during the Second World War.

The next significant development in cutting tool materials was the introduction of cemented carbides by Schroter (2) in 1923. The very high hardness and abrasion resistance of hard metal tools based on tungsten carbide allowed higher metal removal rates under rigidly controlled cutting conditions. The introduction of ceramics, cast cobalt alloys and diamond for specialised applications further extended the range of cutting tool materials. In recent years other significant developments have included improved grades of cemented carbides and improved manufacturing methods for high-speed steels including powder metallurgy and spray-deposited production. Hard coatings on both cemented carbides and high-speed steels and titanium nitride coating of high-speed steels promise further improvements in cutting tool performance. Recent developments of the silicon, aluminium oxynitrides show promise as ceramic cutting tools. A liquid phase sintering mechanism leads to high densification which gives good strength properties. Further, a low expansion coefficient leads to good thermal shock resistance.

4.2. REQUIREMENTS OF TOOL MATERIALS

Whilst the ideal requirements of a satisfactory cutting tool are easily defined it is more difficult to specify a material which has these requirements over a wide range of cutting conditions. Under specified conditions a particular cutting tool material may perform very well whilst it may perform badly if conditions are changed. For example, a ceramic tool will perform well at high cutting temperatures if it is rigidly held and cuts continuously whereas it would be liable to fail in a brittle manner if used for low-speed interrupted cutting involving shock loading. As a general rule the cutting performance of a tool material increases with an increase in hardness. It is, however, well known that there are many exceptions to this general trend, e.g. the most successful hard material used in hard metal carbides, tungsten carbide, has a lower hardness (1600 $kg\,mm^{-2}$) than titanium carbide (2400 $kg\,mm^{-2}$).* Titanium carbide is cheaper and more abundant than tungsten carbide but, nevertheless, tungsten carbide forms the basic material of the present hard metal industry because in combination with the binder phase cobalt it provides the best generally available cutting properties. Attempts to substitute nickel and iron for the binder phase cobalt and substitutions of titanium carbide, tantalum carbide and other transition metal carbides

*Hardness data from M. C. Shaw (1965). *Metal cutting principles*, 3rd Ed. MIT Press, Cambridge, Mass.

for tungsten carbide have not produced significant improvements over the basic grade.

Materials used to cut metals and alloys include carbon steels, high-speed steels, cast cobalt alloys, cemented carbides, ceramics and diamond. The physical and metallurgical requirements of a good cutting tool material include the following:

(1) High yield strength at cutting temperatures;
(2) High fracture toughness;
(3) High wear resistance;
(4) High fatigue resistance;
(5) High thermal capacity and high thermal conductivity;
(6) Low solubility in the workpiece material;
(7) High thermal shock resistance; and
(8) Good oxidation resistance.

The hardness of a material is perhaps the most readily available indicator of its yield strength. The hardness test consists of loading a diamond indentor and pressing it into the material being tested. The further the indentor presses into the material the softer the material and the lower its yield strength. The true hardness (H) is defined as the load (F) divided by the projected area (A) of the indentation (F/A). The yield strength σ_y of the material is related to the hardness since $H=3\sigma_y$. A correction factor is required for materials which work-harden appreciably.

The yield strengths of a number of materials used as cutting tool materials, either alone or in combination, are given in Table 4.2.

Tool materials should resist deformation and fracture during cutting. Materials having high yield strengths are therefore preferred. Unfortunately materials having high strengths often have low toughness and low fracture toughness. The toughness of a material, G_c, is the energy absorbed in making a unit area of crack. A material such as copper has a high toughness ($G_c \simeq 10^6$ J m^{-2}) which means that a large amount of energy is required to make a crack propagate. Alumina, on the other hand, has a low toughness ($G_c \simeq 20$ J m^{-2}).

The condition for the onset of fast fracture in materials is given by

$$\sigma\sqrt{\pi a} = \sqrt{EG_c} \quad [4.1]$$

where the left hand side of the equation implies that fast fracture will occur when, in a material subjected to a stress σ, a crack reaches some

TABLE 4.2
THE YIELD STRENGTH OF A RANGE OF MATERIALS USED IN CUTTING TOOLS (3)

Material	*Yield strength* σ_y ($MN\,m^{-2}$)
Diamond	50 000
Silicon carbide (SiC)	10 000
Silicon nitride (Si_3N_4)	8 000
Tungsten carbide (WC)	6 000
Niobium carbide (NbC)	6 000
Alumina (Al_2O_3)	5 000
Beryllia (BeO)	4 000
Titanium carbide (TiC)	4 000
Zirconium carbide (ZrC)	4 000
Tantalum carbide (TaC)	4 000
Zirconia (ZrO_2)	4 000
Cobalt and alloys	180–2 000
Low alloy steels, quenched and tempered	500–1 980
Carbon steels, quenched and tempered	260–1 300

critical size a or alternatively the material will fracture when the material containing cracks of size a is subjected to some critical stress σ. E is the Young's modulus of elasticity. The critical combination of crack length and stress at which fast fracture occurs is a material constant. The quantity $\sigma\sqrt{\pi a}$ is usually abbreviated to the symbol K, called the stress intensity factor. Fast fracture occurs when $K = K_c$, where $K_c = \sqrt{EG_c}$ (the fracture toughness). The units of fracture toughness are $MN\,m^{-3/2}$. The fracture toughness of a material is measured by inserting a crack of length a and measuring the stress which causes fast cracking. The toughness G_c can be determined directly from the fracture toughness K_c since $K_c = \sqrt{EG_c}$. Toughness and fracture toughness data are tabulated in Table 4.3 for a range of materials including tool materials.

Comparison of the yield strength and fracture toughness data of Tables 4.2 and 4.3 shows that fracture toughness falls with increasing yield strength. The ideal combination then, for a metal cutting tool, of high fracture toughness and high yield strength cannot be achieved in practice. Practical tool materials are usually manufactured to give the maximum possible strength with acceptable fracture toughness.

Cutting tool materials can be manufactured to different strength (hardness) levels. Recent data collected by Almond (4) (Fig. 4.1) shows

TABLE 4.3
TOUGHNESS G_c AND FRACTURE TOUGHNESS K_c (3)

Material	$G_c(\mathrm{kJ\,m^{-2}})$	$K_c(\mathrm{MN\,m^{-3/2}})$
Pure ductile metals (e.g. Cu, Ag, Ni, Al)	100–1 000	100–350
High strength steels	15–118	50–154
WC–Co cermets	0·3–0·5	14–16
Silicon nitride (Si_3N_4)	0·1	4–5
Silicon carbide (SiC)	0·05	3
Alumina	0·02	3–5

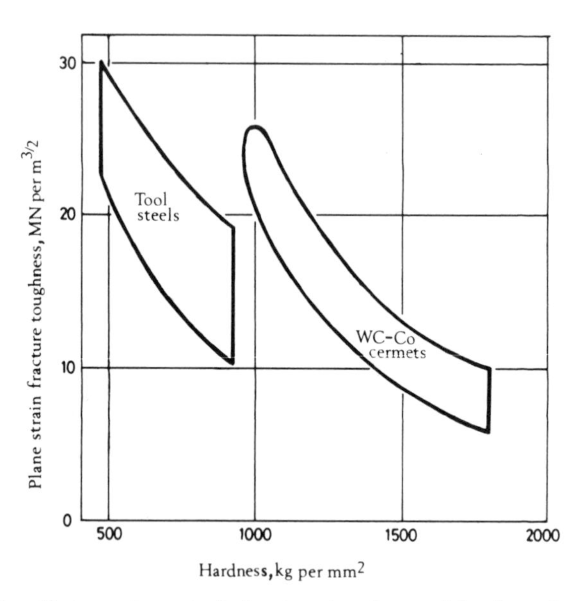

FIG. 4.1. The effect on plane strain fracture toughness of hardness for tool steels and WC–Co cermets (after Almond (4)).

that for tool steels the plane strain fracture toughness K_{Ic} falls with an increase in hardness and that high hardness must require a sacrifice in toughness. Similar results are shown for WC–Co cermets whose fracture toughness falls to low levels as the hardness is raised.

The usefulness of metal cutting tool materials depends to a large extent on their ability to retain good mechanical strength at the elevated temperatures generated in metal cutting. The effect of temperature on hardness (and hence yield strength) of cast cobalt alloys, high-speed steels, WC–Co cemented carbides, WC–TiC–TaC–Co cemented carbides and alumina are shown in Fig. 4.2 (after Almond (4)). The rapid fall-off in hardness of high-speed steels above about 600°C limits their use to temperatures below 600°C. The retention of strength to higher temperatures of the cemented carbides and ceramics renders them preferable for machining at high cutting speeds and for machining difficult-to-machine alloys during which high temperatures are generated.

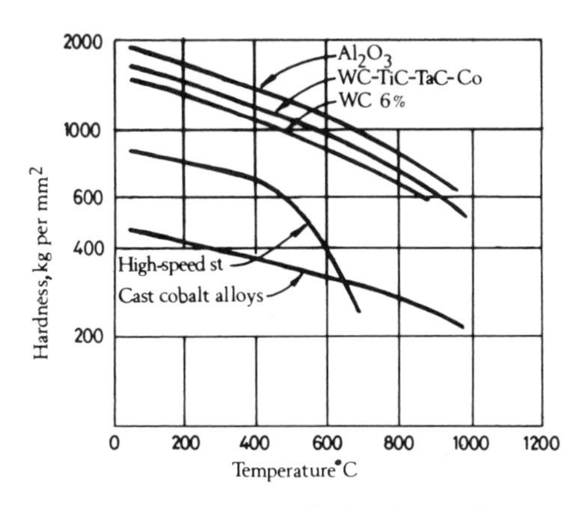

FIG. 4.2. The effect of temperature on the hardness of various tool materials (after Almond (4)).

4.3. HIGH-SPEED STEELS

4.3.1. Introduction

High-speed steels were developed at the end of the 19th century from air-hardening alloy steels. F. W. Taylor and M. White of the Bethlehem Iron Company discovered in 1898 that a high temperature heat-treatment of a wide range of alloy steels allowed these steels to cut at much higher cutting speeds than those previously used. Becker (5) showed in 1910 that significant increases in maximum permitted cutting speed could be achieved by using properly heat-treated high-speed steel rather than carbon tool steels or the original Taylor–White high-speed steel containing 1·85%C, 4%Cr and 8%W. The considerable increase in permitted cutting speed for steels, cemented carbides and ceramics is shown in Fig. 4.3. High-speed steels are manufactured by normal manufacturing routes, i.e. steelmaking, forging, rolling, and machining. In

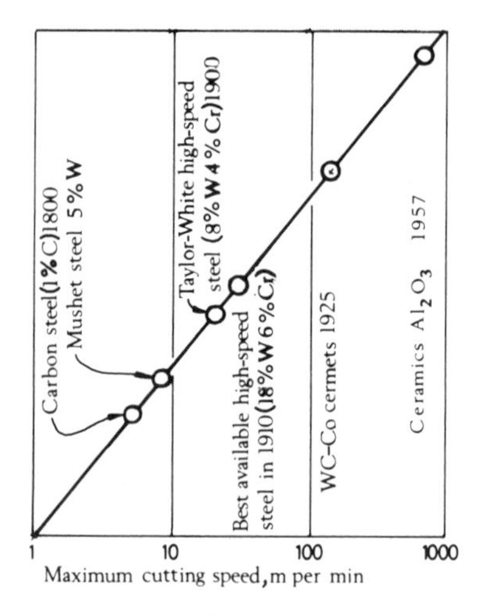

FIG. 4.3. The increase in maximum allowed cutting speed with the introduction of new tool materials.

recent years powder metallurgy techniques have been applied to the manufacture of high-speed steels resulting in a more uniform structure without carbide segregation. These techniques have also allowed the manufacture of more highly alloyed compositions which could not be manufactured by conventional routes due to poor hot workability. Recently, controlled spray-deposited high-speed steels have been produced in which the normal powder production and compaction stages have been eliminated; the fineness of the structure and the absence of carbide segregation is claimed to give a superior product.

High-speed steels retain sufficient hardness at the cutting edge of the tool to cut metal at the rapid rates which generate high temperatures (up to 600°C). High-speed steels are still sold under manufacturers trade names although certain analyses are accepted as standard. The main high-speed steel classifications accepted by AISI, SAE and BSI are the M types and T types containing principally molybdenum and tungsten, respectively. These are further subdivided as shown in Table 4.4 (6). The characteristics described include wear resistance, hardness retention at elevated temperature, safety in hardening, and grindability; also tabulated in Table 4.5 are working and heat-treatment details together with the working hardness of the material.

The principal reason for the good wear resistance of high-speed steel is the retention of high hardness at cutting temperatures. Payson (7) has correlated room temperature and elevated temperature hardness for T1 and M2 high-speed steels. The room temperature and elevated temperature hardnesses are plotted in Fig. 4.4. These curves show an almost linear relation between the two hardness values and a high room temperature hardness results in a high elevated temperature hardness.

TABLE 4.4
CLASSIFICATION OF HIGH-SPEED STEELS (6)

Type	*C(%)*	*W(%)*	*Mo(%)*	*Cr(%)*	*V(%)*	*Co(%)*
Molybdenum types						
M1	0·80	1·50	8·00	4·00	1·00	
M2	0·85; 1·00	6·00	5·00	4·00	2·00	
M3 Class 1	1·05	6·00	5·00	4·00	2·40	
M3 Class 2	1·20	6·00	5·00	4·00	3·00	
M4	1·30	5·50	4·50	4·00	4·00	
M6	0·80	4·00	5·00	4·00	1·50	12·00
M7	1·00	1·75	8·75	4·00	2·00	
M10	0·85; 1·00	—	8·00	4·00	2·00	
M15	1·5	6·5	3·5	4·00	5·00	
M30	0·80	2·00	8·00	4·00	1·25	5·00
M33	0·90	1·50	9·50	4·00	1·15	8·00
M34	0·90	2·00	8·00	4·00	2·00	8·00
M36	0·80	6·00	5·00	4·00	2·00	8·00
M41	1·10	6·75	3·75	4·25	2·00	5·00
M42	1·10	1·50	9·50	3·75	1·15	8·00
M43	1·20	2·75	8·00	3·75	1·60	8·25
M44	1·15	5·25	6·25	4·25	2·25	12·00
M46	1·25	2·00	8·25	4·00	3·20	8·25
M47	1·10	1·50	9·50	3·75	1·25	5·00
Tungsten types						
T1	0·75	18·00	—	4·00	1·00	
T2	0·80	18·00	—	4·00	2·00	
T4	0·75	18·00	—	4·00	1·00	5·00
T5	0·80	18·00	—	4·00	2·00	8·00
T6	0·80	20·00	—	4·50	1·50	12·00
T8	0·75	14·00	—	4·00	2·00	5·00
T15	1·50	12·00	—	4·00	5·00	5·00

On a theoretical basis it can readily be shown for both adhesive and abrasive wear mechanisms that the wear rate of a material per unit sliding distance is proportional to the applied normal load and inversely

TABLE 4.5
CHARACTERISTICS, PROPERTIES AND HEAT TREATMENT DETAILS OF HIGH-SPEED STEELS (6)

AISE–SAE designation	*Characteristics*			
	Performance		*Safety in hardening*[a]	*Ease of grinding*[a]
	Wear resistance[a]	*Hardness retention at elevated temperature*[a]		
T1, T2	VG	VG	G	F
T4	VG	B	F	F
T5	VG	B	F	F
T6	VG	B	F	F
T8	VG	B	F	F
T15	B	B	F	P
M1	VG	VG	F	F
M2	VG	VG	F	F
M3	VG	VG	F	P
M4	B	VG	F	P
M6	VG	VG	F	F
M7	VG	VG	F	F
M10	VG	VG	F	F
M30	VG	B	F	F
M33	VG	B	F	F
M34	VG	B	F	F
M36	VG	B	F	F
M41	VG	B	F	F
M42	VG	B	F	F
M43	VG	B	F	F
M44	VG	B	F	F
M46	VG	B	F	F
M47	VG	B	F	F

[a] Ratings: P = poor, F = fair, G = good, VG = very good, B = best.

proportional to the yield stress (or hardness) of the wearing material (σ_T or H_T). This can be written as

$$\text{Wear rate of tool material} = \frac{K \times \text{load}}{H_T} \qquad [4.2]$$

The constant K for an adhesive wear mechanism is the probability of a wear particle being formed and is a measure of the ratio of the volume of metal removed to the volume of metal ploughed by the abrading particles in abrasive wear. These theoretical models are also supported

TABLE 4.5—*Contd.*

Heat-treatment								
Forging and annealing				*Hardening and tempering*				
Forging range (start to finish) (°C)	*Annealing range* (°C)	*Maximum cooling rate for anneal* (°C hr^{-1})	*Annealed hardness (BHN)*	*Hardening temperature range* (°C)	*Resistance to decarburisation*[a]	*Quenching medium*[b]	*Usual tempering range* (°C)	*Usual working hardness* (R_C)
1120–950	870–900	22	217–255	1260–1300	G	A,S or O	540–590	66–60
1120–950	870–900	22	228–269	1260–1300	F	A,S or O	540–590	66–62
1120–980	870–900	22	235–277	1275–1315	P	A,S or O	540–590	65–60
1120–980	870–900	22	248–293	1275–1315	P	A,S or O	540–590	65–60
1120–950	870–900	22	228–255	1260–1300	F	A,S or O	540–590	65–60
1120–980	870–900	22	241–277	1205–1260	F	A,S or O	540–650	68–63
1090–925	830–870	22	207–235	1175–1220	P	A,S or O	540–590	65–60
1090–925	840–885	22	212–241	1190–1230	F	A,S or O	540–590	65–60
1090–925	840–885	22	223–255	1205–1230	F	A,S or O	540–590	66–61
1090–925	870–900	22	223–255	1205–1230	F	A,S or O	540–590	66–61
1090–925	870–900	22	248–277	1175–1205	P	A,S or O	540–590	66–61
1090–925	830–870	22	217–255	1175–1225	P	A,S or O	540–590	66–61
1090–925	830–870	22	207–235	1175–1220	P	A,S or O	540–590	65–60
1090–925	870–900	22	235–269	1205–1230	P	A,S or O	540–590	65–60
1090–925	870–900	22	235–269	1205–1230	P	A,S or O	540–590	65–60
1090–925	870–900	22	235–269	1205–1230	P	A,S or O	540–590	65–60
1090–925	870–900	22	235–269	1220–1245	P	A,S or O	540–590	65–60
1090–925	870–900	22	235–269	1190–1215	P	A,S or O	540–590	70–65
1090–925	870–900	22	235–269	1190–1210	P	A,S or O	510–590	70–65
1090–925	870–900	22	248–269	1190–1215	P	A,S or O	510–590	70–65
1090–925	870–900	22	248–293	1200–1225	P	A,S or O	540–625	70–62
1090–925	870–900	22	235–269	1190–1220	P	A,S or O	525–565	69–67
1090–925	870–900	27	235–269	1175–1205	P	A,S or O	525–590	70–65

[b] Quenching mediums (in order of decreasing severity): O = oil, S = salt, A = air.

by experimental evidence. It follows, therefore, that the wear rate of tool materials will be reduced by increasing the hardness of the tool material. In practice, as has been described, fracture toughness falls with increasing hardness so that a compromise hardness should be chosen which allows the tool material to retain adequate fracture toughness, whilst at the same time inhibiting abrasive wear.

4.3.2. Structure of High-Speed Steels

In the annealed condition high-speed steels consist of alloy carbides in a lightly alloyed ferritic matrix. The main carbides in high-speed steel in

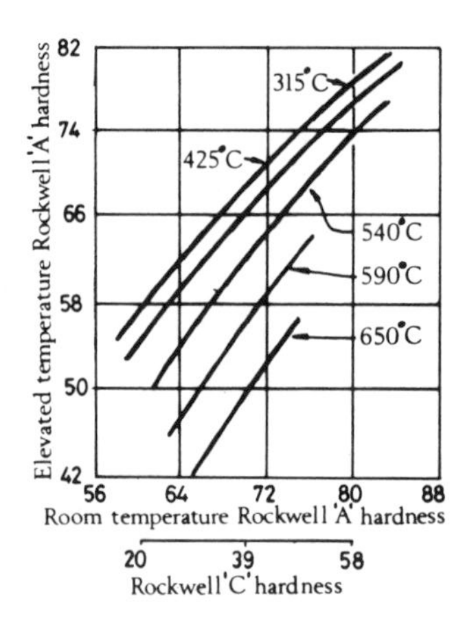

FIG. 4.4. Relationship between room temperature and elevated temperature hardness for M1 and T1 high-speed steels (after Payson (7)).

the annealed condition are the chromium-rich carbide $M_{23}C_6$, the tungsten molybdenum iron double carbide M_6C and the vanadium-rich carbides MC/M_4C_3 (7). Kayser & Cohen (8) measured the quantities of these carbides in a range of high-speed steels in both the annealed condition and the condition following heating up to the hardening temperature. The results are shown in the bar chart of Fig. 4.5. The chromium-rich $M_{23}C_6$ carbide is completely dissolved in all the steels but the tungsten- and molybdenum-rich double carbide M_6C and the vanadium-rich carbides MC/M_4C_3 are only partially dissolved. The excellent red hardness of high-speed steels is due to the highly alloyed austenite formed prior to hardening. The effect of the austenitising temperature on the distribution of alloying elements between the matrix and the carbides in M1 and M4 high-speed steels has also been studied by Kayser & Cohen (8)—the results are shown in Fig. 4.6. It is important to note that the matrix is enriched in both carbon and alloying elements as the austenitising temperature is raised. Whilst most of the chromium is dissolved much less of the other alloying elements—carbon, molybdenum, tungsten and vanadium—are dissolved in the austenite. The remainder of these elements are in the form of residual carbides which inhibit grain coarsening of the austenite, remain unchanged in the final tempered martensitic microstructure and additionally provide abrasion resistance to the finally hardened high-speed steel.

In the heating for hardening of high-speed steel a fine austenitic grain

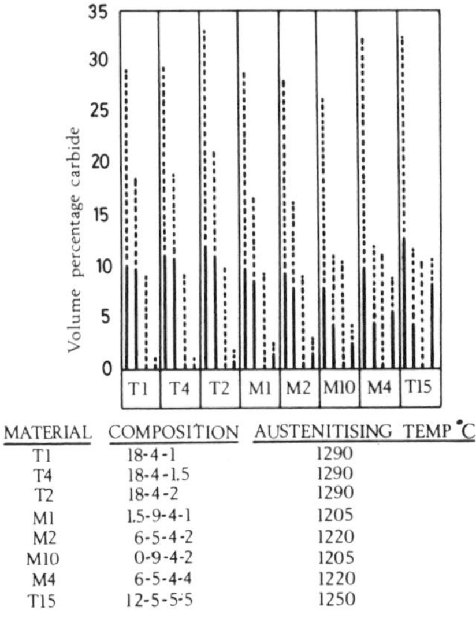

MATERIAL	COMPOSITION	AUSTENITISING TEMP °C
T1	18-4-1	1290
T4	18-4-1.5	1290
T2	18-4-2	1290
M1	1.5-9-4-1	1205
M2	6-5-4-2	1220
M10	0-9-4-2	1205
M4	6-5-4-4	1220
T15	12-5-5·5	1250

Line 1-Total Line 2-M_6C Line 3-$M_{23}C_6$ Line 4-MC

Solid lines – After heating to austenitising temperature
Broken lines – In the annealed condition

FIG. 4.5. The quantities of the main carbides present in a number of high-speed steels in the annealed condition and after heating to the austenitising temperature (after Kayser & Cohen (8)).

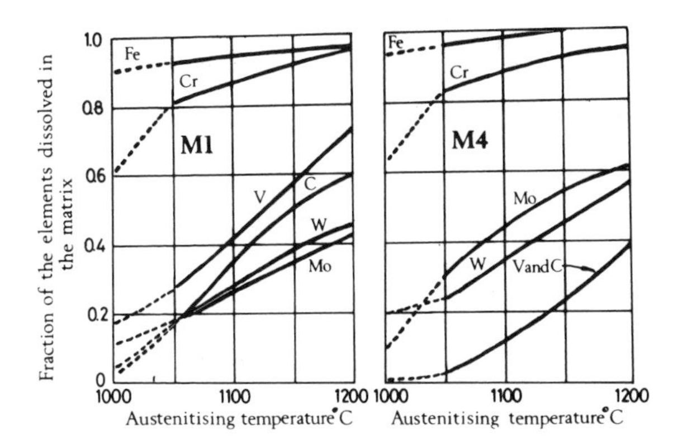

FIG. 4.6. Distribution of elements between matrix and carbides in M1 and M4 high-speed steels (after Kayser & Cohen (8)).

size is important; excessive grain growth at austenitising temperatures causes reduced toughness of the high-speed steel following quenching and tempering. Grobe *et al.*(9) measured austenitic grain size as a function of austenitising temperature for four high-speed steels using the Snyder–Graff intercept method. The effect of austenitising temperature on the Snyder–Graff intercept grain size is shown in Fig. 4.7.

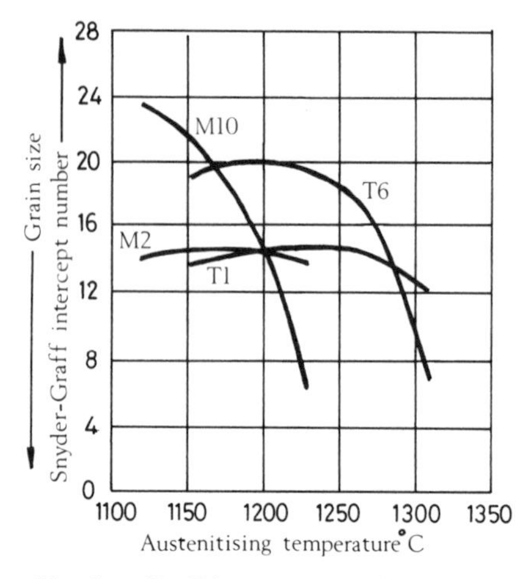

FIG. 4.7. Grain size (Snyder–Graff intercept number) of several high-speed steels as a function of austenitising temperature (after Grobe *et al.* (9)).

The importance of M2 and T1 high-speed steels and closely related members of their families is due, in part, to the resistance of these steels to grain coarsening at austenitising temperatures. This behaviour is in contrast to that of M10 and T6 high-speed steels which show a strong temperature-dependence of grain coarsening rate. The latter high-speed steels, therefore, require much closer control of heat treatment than the more popular M2 and T1 high-speed steels.

4.3.3. Heat Treatment of High-Speed Steels

High-speed steels are normally austenitised followed by oil quenching and tempering. The structure of quenched high-speed steel consists of primary carbides (those which were not dissolved on austenitising) in a matrix of martensite and austenite. Using optical metallography it is difficult to distinguish austenite from martensite in a quenched high-speed steel. The use of X-ray and magnetic techniques have shown that normally hardened high-speed steels contain between 15 and 20% retained austenite.

4.3.3.1. Equilibrium Diagram of High-Speed Steels

Phase changes in high-speed steels which are complex alloys cannot be represented by conventional equilibrium diagrams. However, it is possible to represent phase changes by using a binary section through a more complex system. For the T1 high-speed steel a binary section through the Fe–W–Cr–V–C system at the 18%W–4%Cr–1%V composition is shown in Fig. 4.8. This diagram is only approximate but

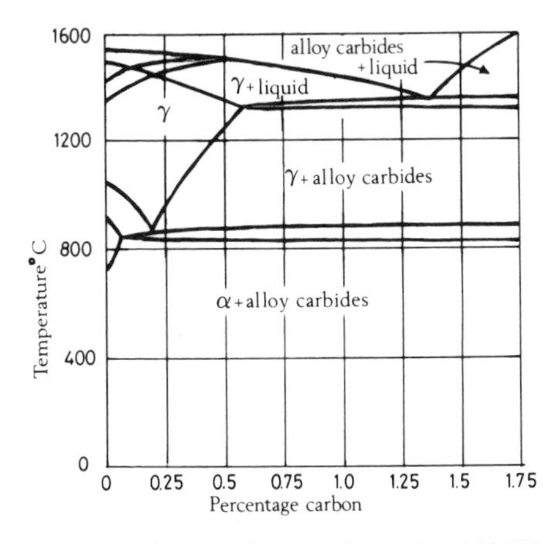

FIG. 4.8. Pseudo-binary equilibrium diagram for an Fe–18%W–4%Cr–1%V alloy with different carbon contents.

is useful in discussing heat-treatment principles. It should be noted from Fig. 4.8 that both eutectic and eutectoid reactions take place over a range of temperatures in high-speed steels. The addition of 18%W–4% Cr–1%V to iron raises the A_1 temperature from 723°C to approximately 850°C; the eutectic temperature is also raised—from 1130°C to 1330°C. The solubility of carbon in austenite is reduced at 1330°C from 1·7% in plain carbon steels to 0·7% in T1 high-speed steel. The pseudo-binary diagram for T1 high-speed steel indicates that it is necessary to heat it to around 1300°C to achieve high solubility of carbon in austenite. The high carbon content of the austenite prior to quenching ensures that a hard martensite is obtained on quenching since the hardness of martensite is approximately directly proportional to its carbon content.

4.3.3.2. Time–Temperature Transformation Diagram

The best way to represent transformation data of a steel is to plot it on a time–temperature graph. The time scale is logarithmic so that short times of the order of a few seconds and long times of the order of many hours can be represented on the same diagram. A time–temperature transformation (TTT) diagram for an M2 high-speed steel is shown in Fig. 4.9(7). It is apparent from this diagram that the steel has high hardenability since it is possible to air cool it to obtain a martensitic

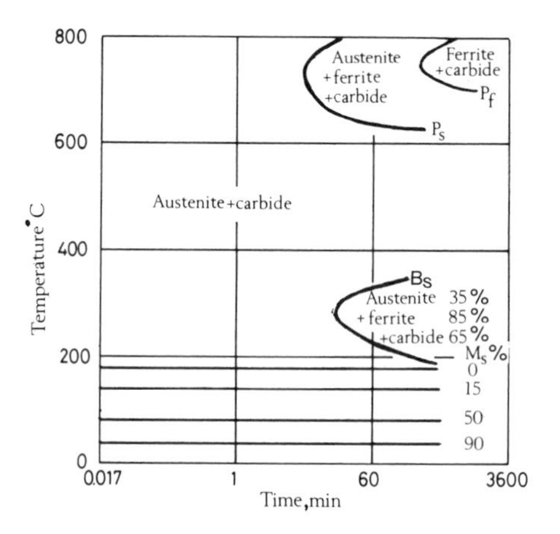

FIG. 4.9. TTT diagram for M2 high-speed steel austenitised at 1230°C.

structure. Further, the transformation in the upper right-hand corner of the diagram to ferrite and carbide (normally designated P_s and P_f in carbon and low alloy steels) is displaced to higher temperatures compared with plain carbon steels. The start of the transformation to bainite (B_s) is shown in the lower right-hand side of the diagram; the extent to which the austenite is transformed to bainite for three different transformation temperatures is given as 35, 85 and 65% after a transformation time of between 15 and 20 h. The transformation to martensite commences at the temperature line designated M_s; the amount of transformation of austenite to martensite increases with continuous fall in temperature below M_s.

4.3.3.3. Tempering of Martensite

The tempering of high-speed steels is carried out to improve mechanical properties by allowing a secondary precipitation-hardening process to occur and retained austenite also transforms during tempering although the transformation of retained austenite is not considered to contribute directly to secondary hardening. The tempering curve for an M2 high-speed steel is shown in Fig. 4.10. The shape of this curve is typical of most tempering curves for high-speed steels which show a pronounced secondary hardening peak. The secondary hardening is considered by recent workers to be due to precipitation of V_4C_3 and possibly the simultaneous precipitation of W_2C(10). During tempering the growth of

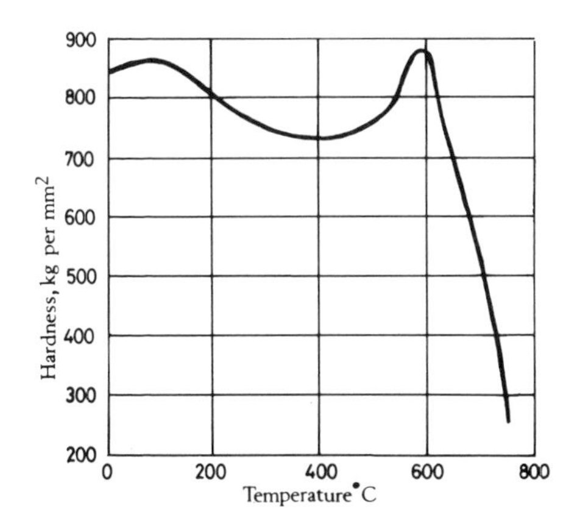

FIG. 4.10. Tempering curve for an M2 high-speed steel showing the effect of holding for an hour at different tempering temperatures on the hardness following hardening from 1170°C.

secondary carbides is controlled by diffusion. The shape of the tempering curve and, in particular, the temperature at which the peak hardness is reached is clearly dependent on tempering time. Like all precipitation-hardening mechanisms the maximum hardness is reached at lower temperatures for long holding times. A graph of a single parameter T $(C + \log t)$ against hardness gives a universal tempering curve for M2 high-speed steel and this curve is shown in Fig. 4.11 where T and t are temperature and time and C is a constant.

4.3.4. Applications of High-Speed Steels

The composition and some important properties of high-speed steels have been summarised in Tables 4.4 and 4.5. For the purpose of discussing the applications of high-speed steels it is convenient to classify them in groups based on their applications. Table 4.6 (after Kirk *et al.* (11)) lists a few of the more common high-speed steels based on normal duty, higher cutting speed applications and applications related to the machining of harder materials such as those used in aerospace

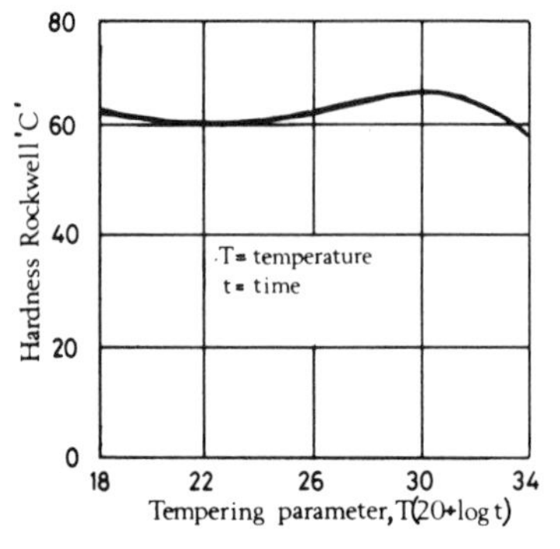

FIG. 4.11. Master tempering curve for M2 high-speed steel (after Payson (7)).

TABLE 4.6
PROPERTIES OF HIGH-SPEED STEELS (11)

Type	*Hardness Rockwell 'C'*	*Relative toughness (J)*	*Red hardness Rockwell 'C' after triple tempering at 650°C*
Normal duty			
M1	63–65	32·5	50
M2	63–65	34	53
T1	63–65	20	55
Higher speeds			
M35	64–66	27	55
T4	64–66	5	57
T6	65–67	11	59
Harder materials			
M15	66–68	22	52
M42	66–70	18	50

applications. Also listed is an empirical indication of the toughness based on unnotched Izod impact tests and the measured hardness after triple tempering at 650°C.

The choice of a high-speed steel within one of the three groupings in Table 4.6 should take account of hardness, toughness, hardness retention at cutting temperatures, grindability, machinability, wear resistance, and the particular application as well as price. The steels in the normal duty group are largely interchangeable, but the T1 grade whilst it has a higher red hardness than the others has a lower toughness which would render it less suitable than M2 for applications where tool failure based on fracture of the cutting edge occurs. The M1 grade has a higher molybdenum content than the M2 grade which makes it more prone to decarburisation during manufacture; however, it has a lower vanadium content which renders it more grindable. Considering the higher speed group of high-speed steels the T6 grade has the highest hardness retention at elevated temperature and is particularly suitable for single point cutting operations where it is not subjected to repeated shock loading which may lead to breakdown of the cutting edge due to fracture processes. The lower alloyed T4 grade is less expensive than T6 and provides a cheaper alternative. The M35 grade retains a lower hardness at elevated temperatures than either the T4 or T6 grades but has improved toughness which allows a wider range of applications. In the machining of harder materials, such as those used in aerospace applications, the primary tool material property is wear resistance; the presence of hard vanadium carbide in increasing quantities improves the abrasion resistance of such steels. The M15 grade containing 5%V has very high abrasion resistance but poor grindability.

The better grindability of the M42 grade has led to its increasing use for machining the harder materials since it also has a high hardness arising principally from a hard matrix strengthened by cobalt. It is important to appreciate that for a given grade of high-speed steel an increase in hardness which gives improved wear resistance also generally leads to a decrease in fracture toughness. For continuous-cutting operations a high hardness may be the most desirable single property whereas in interrupted cutting processes a higher toughness with a corresponding lower hardness may be preferable for improved tool performance.

4.4. CEMENTED CARBIDES

4.4.1. Introduction

Cemented carbides (or cermets) are mixtures of transition metal carbides and metals in which the metal, usually cobalt, binds the hard carbides together. Cemented carbides are manufactured by powder metallurgical techniques involving production of the hard carbide particles, compaction and sintering at high temperatures. Cemented carbides are produced in a wide variety of compositions with varying properties to meet a wide range of applications. The bulk of the cemented carbides in use today have tungsten carbide (WC) as the primary component with cobalt as the binder metal. Together with high-speed steels they form the two important groups of cutting tool materials.

4.4.2. Classification of Cemented Carbides

Cemented carbides have been classified by the International Standards Organisation (ISO) on the basis of the material to be machined. This classification is given in Table 4.7. An alternative classification based on the machining application has been agreed by the Joint Industrial Council of the United States of America and is given in Table 4.8. In the ISO classification, cutting speed and wear resistance increase from the bottom to the top of the classification whilst toughness and feed rate increase from top to bottom of the classification.

TABLE 4.7
ISO CLASSIFICATION OF CEMENTED CARBIDES BASED ON MATERIAL TO BE MACHINED[a]

Main machining group	*Application group*	*Operations and working conditions*
P: steel, cast steel.	P01	High precision turning and boring, high cutting speeds, small chip cross-section, dimensional accuracy, good surface finish, and vibration-free machining.
	P10/20	Turning, copy turning, thread cutting and milling, high cutting speeds, and small to medium chip cross-section.
	P30	Turning, milling, planing, medium to low cutting speeds, medium to large chip cross-section, also under favourable conditions.
	P40	Turning, planing, milling, shaping, low cutting speeds, large chip cross-section, high rake angles, unfavourable conditions; also automatic turning.
	P50	Where highest demands are made on toughness of carbide: turning, planing and shaping, low cutting speeds, large chip cross-section, and high rakes under unfavourable conditions. Automatic turning.

TABLE 4.7—*contd.*

Main machining group	*Application group*	*Operations and working conditions*
M: steel, cast steel, austenitic manganese steel, cast-iron alloys, austenitic steels, malleable and spheroidal cast iron, free cutting mild steel	M10	Turning, medium to high cutting speeds, small to medium chip cross-section.
	M20	Turning, milling, medium cutting speeds, and medium chip cross-section.
	M30	Turning, milling, planing, medium cutting speeds, medium to large chip cross-section.
	M40	Turning, form turning, parting off and recessing, particularly for automatics.
K: cast iron, chilled cast iron, hardened steel, non-ferrous metals, non-metallic materials	K01	Turning, precision turning and precision boring, finish milling, and scraping.
	K10	Turning, milling, boring, countersinking, reaming, scraping and broaching.
	K20	Turning, milling, planing, countersinking, scraping, reaming, and broaching under tougher conditions than K10.
	K30	Turning, milling, planing, shaping under unfavourable conditions, high rakes.
	K40	Turning, milling, planing, shaping under unfavourable conditions, high rakes.

[a] Cutting speed and wear resistance increase from bottom to top, feed and carbide toughness from top to bottom.

4.4.3. Structure and Properties of Cemented Carbides

Commercial cemented carbides contain the carbides of tungsten, titanium and tantalum, and those used for metal cutting usually contain at least 80% by volume of carbide. The variation in the hardness of the monocarbides of tungsten, tantalum, titanium and niobium with temperature up to 1200°C is shown in Fig. 4.12 (12, 13). These carbides remain much harder than high-speed steels over a very wide temperature range even though their hardness falls rapidly with rise in temperature.

TABLE 4.8
JIC CARBIDE-CLASSIFICATION CODE (2)[a]

Code	*Application*	*Carbide characteristics*
C–1	Roughing	Medium–high shock resistance Medium–low wear resistance
C–2	General-purpose	Medium shock resistance Medium wear resistance
C–3	Finishing	Medium–low shock resistance Medium–high wear resistance
C–4	Precision finishing	Low shock resistance High wear resistance
C–5	Roughing	Excellent resistance to cutting temperature Shock and cutting load Medium wear resistance
C–50	Roughing and heavy feeds	Same as above
C–6	General-purpose	Medium–high shock resistance Medium wear resistance
C–7	Finishing	Medium shock resistance Medium wear resistance
C–70	Semi-finishing and finishing	High cutting temperature resistance Medium wear resistance
C–8	Precision finishing	Very high wear resistance Low shock resistance

[a] Toughness increases from bottom to top, hardness from top to bottom.

In the plain WC–Co cemented carbides both the cobalt content and the WC particle size have a significant influence on the mechanical properties. The effect of WC particle size on hardness is shown in Fig. 4.13 where it can be seen that hardness increases with a decrease in particle size. Trent (14) has considered in detail the structure and properties of WC–Co alloys and the effect of increasing cobalt content for a medium–fine grain size carbide on the hardness, transverse rupture strength and compressive strength is shown in Fig. 4.14. It is important to note that strength and hardness are highest for low levels of cobalt, the transverse rupture strength, an indication of toughness, increases with increasing cobalt content whilst compressive strength decreases with increasing cobalt content. The manufacturer therefore has a measure of control over strength and toughness by choice of carbide size and cobalt content. The increasing transverse rupture strength with increasing cobalt content

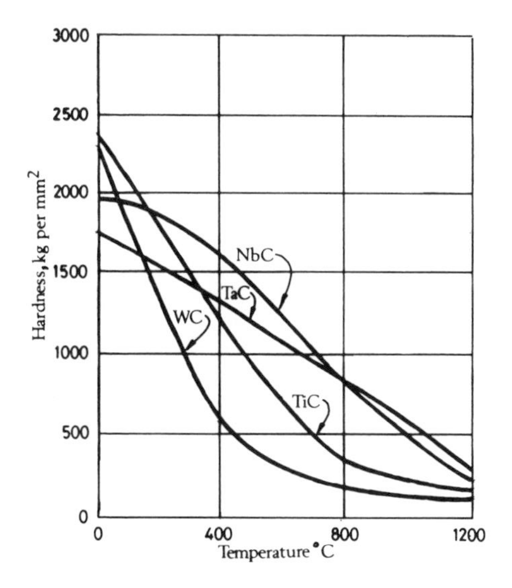

FIG. 4.12. Variation of hardness with temperature for the transition metal carbides of W, Ti, Nb and Ta.

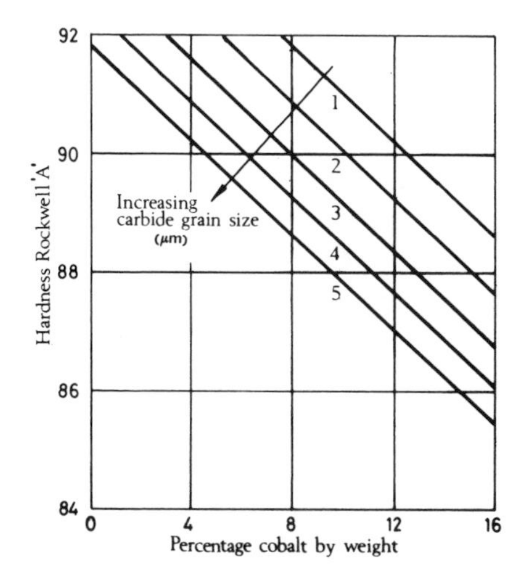

FIG. 4.13. The hardness of WC–Co cemented carbide as a function of cobalt content and WC particle size.

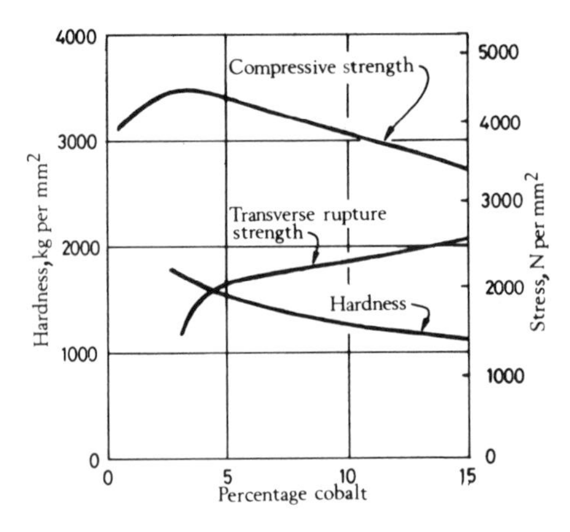

FIG. 4.14. Effect of cobalt content on the mechanical properties of WC–Co alloys (after Trent (14)).

is due to the increasing thickness of the binder phase between the carbide particles. It is interesting to compare the hardness of tungsten carbide and tungsten carbide alloyed with 6 and 9% cobalt (Fig. 4.15). The cemented carbide has significantly higher hardness than the tungsten carbide at temperatures above approximately 200°C. This behaviour, in which the alloying of the hard carbide with a soft matrix material results in improved resistance to deformation at elevated temperatures of the

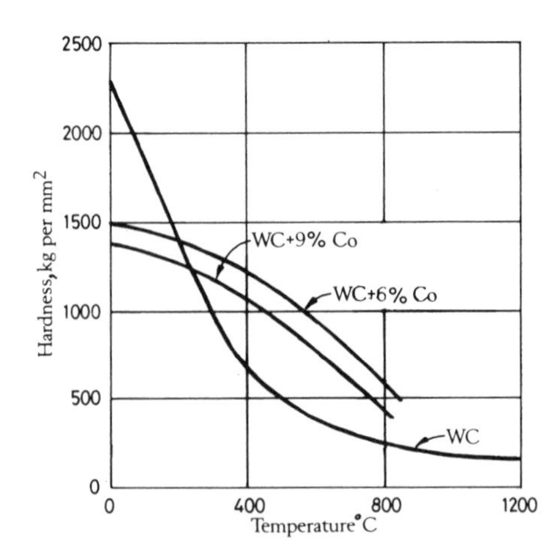

FIG. 4.15. Comparison of the hardness of WC and WC alloyed with 6 and 9% Co as a function of temperature.

alloy over the hard carbides alone, accounts for the remarkable success of this family of materials for cutting tools. Gurland (15) has shown that the strength of WC–Co cemented carbides depends on the ability of the cobalt binder to deform plastically. The brittleness at low cobalt levels and small particle sizes is caused by internal stress conditions which suppress plastic flow. In cemented carbides the plastic flow of the interparticle cobalt layer is inhibited by the rigid skeletal structure of the tungsten carbide. The local stress condition in the cobalt film is considered to be similar to that which exists in a thin soldered joint in steel where the ductile joining metal is prevented from yielding at its bulk yield strength by restraint of the elastically loaded steel. The treatments to date are based on simple and idealised models using continuum mechanics on a microscale. These models calculate a constraint factor which is proportional to the thickness of the cobalt layer. Further work is required to establish quantitative relationships between flow stress and structural parameters. The impact strength of WC–Co cemented carbides increases with an increase in cobalt content, as does the compressive strength up to 4% cobalt. At this level of cobalt the latter then decreases with an increase in cobalt content. Cemented carbides have very high moduli of elasticity which give high resistance against elastic deflection to cemented carbide tools during cutting.

To achieve optimum mechanical properties in cemented carbides it is essential that the product should have high density with freedom from porosity and freedom from the eta phase (Co_3W_3C). The toughness and wear resistance are very much dependent on carbide size and cobalt content. The toughness can be increased by increasing cobalt content, and wear resistance can be increased by lowering the cobalt content or using a carbide having smaller grain size.

4.4.4. Mixed Cemented Carbides Bonded with Cobalt

The WC–Co cemented carbides are highly successful for machining non-ferrous alloys and cast irons but undergo severe crater wear when machining steel. The addition of other carbides including TaC, TiC and NbC increases the cutting speed at which crater wear of the tool takes place. This type of wear is caused by a diffusion mechanism which occurs particularly when long continuous chips are formed. Cemented carbides are therefore divided into two main classes, the straight WC–Co grades for machining cast irons and non-ferrous alloys, and the steel cutting grades which are WC–Co grades containing additions of titanium, tantalum and niobium carbide either alone or in combination. The addition of TaC and TiC to WC–Co cemented carbides improves wear resistance and hot hardness but generally lowers strength. In particular, however, these additions reduce diffusion wear rates on the face of the cutting tool and so allow machining at much higher cutting speeds than the straight grades. The bond strength between steel and the TiC alloyed cemented carbides is much lower than that which forms between steel and the straight grade and this means that during cutting there is less tendency for the moving chip to remove fragments of the tool. Under conditions where the bond is weak the interface will shear leaving the tool unchanged. Trent (14) has provided a satisfactory explanation for the influence of TiC and TaC on reducing cratering of alloyed cemented carbides. He has shown metallographically that the cubic solid solution carbides of WC dissolved in TiC are more resistant to diffusion wear than the hexagonal WC grains. Even small amounts of TiC have a significant influence on reducing crater wear. Increasing the amounts of TiC and TaC improves resistance to cratering wear but also causes a reduction in transverse rupture strength (Fig. 4.16). TaC does not have

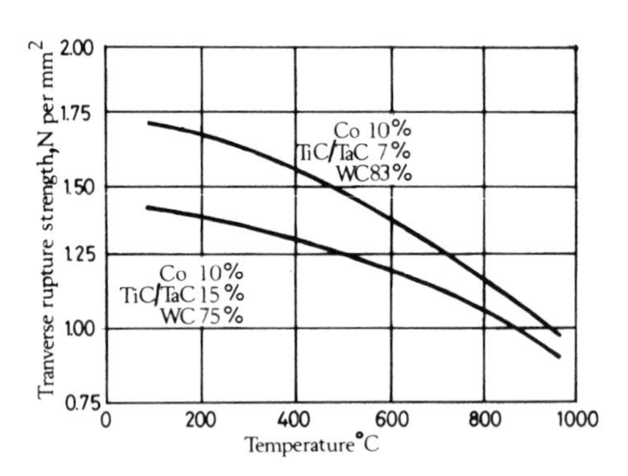

FIG. 4.16. Influence of the level of TiC/TaC on the transverse rupture strength of cemented carbides.

such an adverse effect on transverse rupture strength as TiC and consequently TaC finds use for some specialised applications. Trent (14) has shown, by the use of machining charts, that the steel cutting grades can be used at speeds three times as high as those possible when using the straight carbide grades. This is illustrated in Fig. 4.17 which shows the machining charts for a WC–Co tool with and without 15% TiC when cutting the same steel. Trent further showed that rapid cratering occurred at a cutting speed of 90 m min^{-1} at a feed rate of 0·25 mm on the

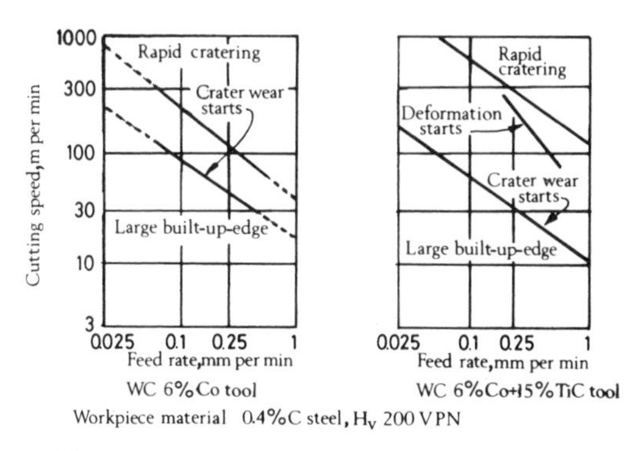

FIG. 4.17. Machining charts for WC–6%Co and WC–6%Co–15%TiC tools showing the combination of cutting speed and feed at which cratering and deformation start (after Trent (14)).

WC–Co tool and at a cutting speed of 270 m min^{-1} for the same feed rate on the tool containing 15% TiC. When steels are machined under cutting conditions where unstable built-up-edges are formed Trent showed that the wear mechanism is one in which the built-up-edge is continuously formed and removed. During this process the alloyed WC–Co cemented carbides perform less well than the straight grades since the cubic carbides are more easily broken up and carried away by the unstable built-up-edge.

4.4.5. Coated Cemented Carbides

In principle it is possible using chemical vapour deposition (CVD) techniques to deposit a wide range of hard compounds onto cemented carbides. Presently both titanium carbide and titanium nitride are being applied commercially, with success, to WC–Co cemented carbides. Such coatings give the improved cratering resistance of the harder TiC and TiN compounds whilst retaining the inherent toughness of the basic WC–Co grade. The TiC coatings are applied by CVD techniques and are of the order of $5–7{\cdot}5 \times 10^{-6}$ m thick. It is claimed that the TiC coating reduces friction between chip and tool and consequently reduces cutting forces by up to 25% and cutting temperatures by up to 100°C. It is also suggested that coated tools are best used on metals which tend to adhere strongly to conventional cemented carbides since the lack of adhesion between workpiece and tool reduces diffusion wear. The major disadvantage of coated tools is that they are suitable only for single application uses, such as indexable inserts, since grinding will remove the coating. Titanium nitride coatings are extremely hard and they act in a similar way to titanium carbide coatings in reducing chip–tool friction with a consequent lowering of tool temperatures and cutting forces and an increase in the cutting speed at which severe cratering starts. It is also thought that at very high cutting speeds titanium oxide is formed at the chip–tool interface and this acts as an efficient barrier to diffusion of the tool into the workpiece. A comparison of the benefits to be obtained from both TiC and TiN coating of a C5 WC–Co cemented carbide is shown in Fig. 4.18.

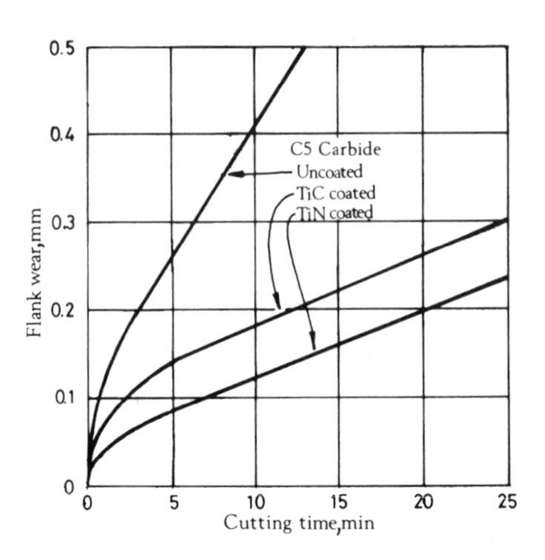

FIG. 4.18. Flank wear–cutting time relationship for uncoated and coated cemented carbides.

4.4.6. Titanium Carbide Cemented Carbides

Tools based on titanium carbide bonded with nickel and molybdenum are gaining in importance. The increasing use of these tools is due to their higher hardness over a wider temperature range than tungsten carbide cemented carbides and the low adhesion of TiC to steel. Titanium carbide grades are claimed to allow higher cutting speeds, give longer tool life, and better surface finish or better tool performance over a wider range of cutting speeds. The better tool performance over a wide range of cutting speeds is due to improved resistance to cratering at high speeds and a lower speed for a given feed for the onset of built-up-edge formation. (The latter leads to high wear rates due to attrition.) Titanium carbide-based tools have proved satisfactory for facing operations where peripheral speeds are too high for high-speed steels and near axial speeds are too low for WC–Co cemented carbides. The potential of TiC-based tools is shown in Fig. 4.19 where they show lower wear rates than WC–Co cemented carbides and also, in one case, a better performance than a TiC-coated WC–Co grade. Titanium carbide-based tools have the further advantage over coated tools since they can be reground.

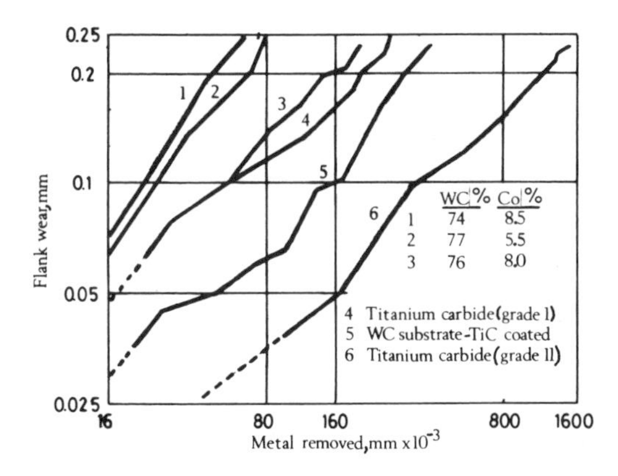

FIG. 4.19. Tool wear behaviour of a range of cemented carbides.

4.5. CAST COBALT ALLOYS

The cast-cobalt-based alloys, known more usually by trade names such as Stellite, find limited use at cutting speeds intermediate between those appropriate for high-speed steels and those appropriate for cemented carbides. These cast alloys were developed during the period 1914 to 1918 and allowed higher cutting speeds than were possible with high-speed steels. A typical composition for a hard grade of one of these alloys is given below:

Carbon	*Chromium*	*Tungsten*	*Manganese*	*Silicon*	*Nickel*	*Cobalt*
2·5%	32%	17%	1% max.	1% max.	2·5% max.	Balance

The alloys are essentially of the C–Cr–W–Co system with minor variations for special applications. The structure of these alloys is complicated and consists of primary trigonal carbides which are acicular in form and are of the Cr_7C_3 type, in which partial substitution of tungsten and cobalt for chromium takes place. The matrix is a complex structure of binary, ternary and higher eutectics of all the constituents of the alloy. The hardness and toughness of the alloys are strongly dependent on their cast structure which can be varied by cooling rate. The effect of temperature on the hardness of cast-cobalt-based alloys is shown in Fig. 4.2. These alloys have a higher hardness above about 600°C than high-speed steels and will consequently give better cutting performance as the temperature is progressively raised above 600°C. The high chromium content of these alloys renders them highly corrosion- and oxidation-resistant and this makes them a natural choice for machining chloride-containing polymers which may give off corrosive products during machining.

4.6. CERAMIC CUTTING TOOL MATERIALS (16)

Ceramic tool materials are dense polycrystalline aggregates of alumina having a grain size of 2 to 5 μm. Small amounts of other ceramics are added to aid sintering and to give a fine structure by limiting grain growth during sintering. The high elevated temperature hardness and wear resistance of alumina are primarily responsible for its use as a cutting tool material. Alumina does not bond strongly to ferrous materials at high temperatures which gives good tool life, by reducing attrition wear, and good surface finish to the machined product. Alumina, however, has low fracture toughness which means that it is not suitable for heavy interrupted cutting. Alumina cutting tools generally require negative rake geometries to increase their strength and reduce the tendency to failure by fracture. The much greater wear resistance of ceramic tool materials compared with cemented carbide and high-speed steel is illustrated by the work of Siekman *et al.* (17). They measured the time required to develop a 1·25 mm flank wear land as a function of cutting speed when machining an AISI 1045 steel. The results are shown in Fig. 4.20. If cutting tools simply required high wear resistance then the ceramic tool would appear at first sight to be far superior. However, tools must also resist fracture and usually wear behaviour alone is not a satisfactory criterion of tool performance.

For ceramic tools a more realistic indication of performance is shown in the work of Kibbey & Moore (18) who measured total tool life in comparing the behaviour of ceramic and cemented carbide tools. The results of their work are shown in Fig. 4.21. Ceramic tools normally fail by fracture; a tool life criterion based on complete tool failure is therefore much more meaningful. The results of Fig. 4.21 show that the cemented

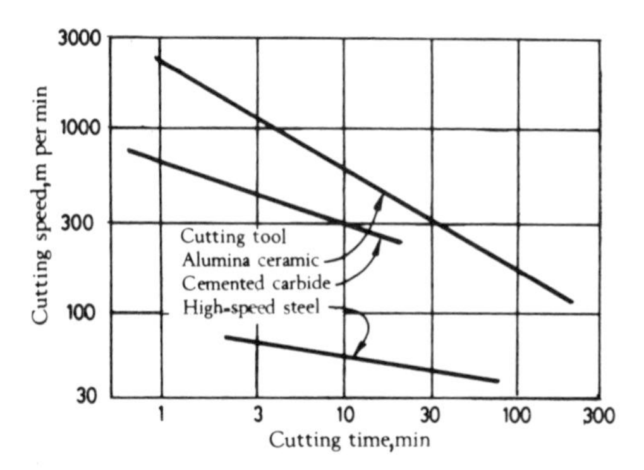

FIG. 4.20. The effect of cutting speed on the time to develop a 1·25 mm flank wear land on ceramic cemented carbide and high-speed steel when machining an AISI 1045 steel.

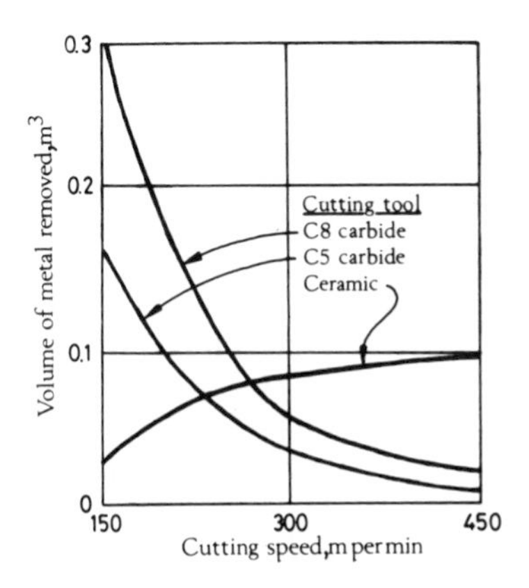

FIG. 4.21. Metal removed versus cutting speed for two carbides and a ceramic.

carbide tool is preferred at speeds below about 240 m min^{-1} since a greater volume of metal can be machined. At higher speeds the ceramic tool performs better. Ceramic tools are therefore preferred for continuous machining of cast irons and hardened steels at high cutting speeds.

Figure 4.2 shows that ceramic tools based on alumina have the highest hardness of all the important tool materials (apart from diamond) over a wide temperature range. On the basis of wear resistance alone, alumina is an extremely attractive cutting tool material. Whilst alumina is chemically inert to many metals, e.g. iron, manganese, cobalt, nickel, vanadium, tantalum, niobium, chromium and molybdenum, it is reduced by magnesium at approximately 900°C when tool wear rates can be extremely rapid. Ceramics are not recommended for machining aluminium and its alloys because of strong bonding between alumina and aluminium. Also a reaction between titanium and zirconium and alumina occurs at approximately 1400°C which causes rapid tool deterioration. Ceramics are also important in the machining of hard cobalt-based alloys and the AISI 400 series stainless steels since cemented carbides chip or wear rapidly with this group of materials. For these materials, and even with ceramic tools, attention must be given to tool design which should include negative rake angles, large lead angles, and large nose radii. The softer work hardening stainless steels and nickel-based alloys have higher shear strengths which require positive tool angles and tools having high fracture toughness; ceramic tools are not recommended for cutting these materials.

4.7. DIAMOND

Diamond is the hardest cutting tool material (hardness $\sim 7000\,\mathrm{kg\,mm^{-2}}$) but has low fracture toughness. Diamond finds use in cutting applications in both single crystal and sintered polycrystalline forms. In single crystal form it is normally lapped to the required shape and mounted in a tool holder and since the wear rate of diamond is anisotropic the diamond should be carefully oriented to give low wear rate on the cutting surfaces of the tool. It is usual to use diamond on non-ferrous materials which are highly abrasive or in which a very good surface finish or high dimensional accuracy is required. An example would be the finish machining of aluminium–silicon alloys containing high silicon levels which are normally highly abrasive to cemented carbides. Diamond in both single crystal and polycrystalline form is recommended for finish machining of abrasive non-ferrous alloys, copper, melamine and phenolic plastics, silica, graphite, cemented tungsten carbide, ceramics, abrasive rubber materials, and fibreglass composites. Polycrystalline diamond inserts are brazed onto the cutting edges of circular saws for cutting construction materials such as concrete and natural stone. Often construction materials contain highly abrasive particles which would give low tool life to most tool materials. Diamonds are not recommended for machining ferrous materials, titanium alloys and high nickel alloys since diamond oxidises or graphitises at the high temperatures generated at the chip–tool interface when machining these materials.

Polycrystalline diamond tools are made by sintering at high pressure (7000 MN m^{-2}) and high temperature (2000°C). A number of forms are available made either solely from sintered diamond or a sintered diamond overlay on a cemented carbide substrate. The use of diamond tools allows higher cutting speeds than those possible with cemented carbides but the deterioration of diamond at only moderately elevated temperatures (at above 600°C it oxidises in air and graphitises in a vacuum) restricts its usefulness to a limited but important range of materials.

REFERENCES

1. Roberts, G. A., Hamaker, J. C. & Johnson, A. R. (1962). *Tool steels*, Third ed., American Society for Metals.
2. Schroter, K. (1923). US Patent No. 1549615.
3. Ashby, M. F. & Jones, D. R. H. (1980). *Engineering materials*, Pergamon Press, London.
4. Almond, E. A. (1982). Towards improved tests based on fundamental properties, In: *Towards improved tool performance*, National Physical Laboratory Conference, Metals Society, London.
5. Becker, O. M. (1910), *High-speed steel*, McGraw-Hill Book Company, New York.
6. Dallas, D. B. (Ed.) (1975). *Tool and manufacturing engineers handbook*, 3rd Ed., Society of Manufacturing Engineers, McGraw-Hill Book Company, New York.
7. Payson, P. (1962). *The metallurgy of tool steels*, J. Wiley & Sons, New York.
8. Kayser, F. & Cohen, M. (1952). Carbides in high-speed steels—their nature and quantity, *Metal Progress*, **61**, 79.
9. Grobe, A. H., Roberts, G. A. & Chambers, D. S. (1954). Discontinuous grain growth in high-speed steels, *Trans. ASM*, **46**, 759.
10. Mukherjee, T. (1970). Physical metallurgy of high-speed steels, In: *Materials for metal cutting*, Iron and Steel Institute, No. 126.
11. Kirk, F. A., Child, H. C., Lowe, E. M. & Wilkins, T. J. (1970). High-speed steel technology—the manufacturers viewpoint, In: *Materials for metal cutting*, Iron and Steel Institute, No. 126.
12. Atkins, A. G. & Tabor, D. (1966). Hardness and deformation properties of solids at very high temperatures, *Proc. Roy. Soc.*, **A 292**, 441–59.
13. Miyoshi, A. & Hara, A. (1965). High temperature hardness of WC, TiC, TaC, NbC and their mixed carbides, *Japan Soc Powder Metallurgy*, **12**, 78–84.
14. Trent, E. M. (1977). *Metal cutting*, Butterworths, London.
15. Gurland, J. (1970). *Microstructural aspects of the strength and hardness of cemented tungsten carbide*, ISI Publication 125.

16. King, A. G. & Wheildon, W. H. (1966). *Ceramics in machining processes*, Academic Press, London & New York.
17. Siekman, H. J., Goliber, E. W. & Stalker, K. W. (1956). Roundup on tomorrows tools materials, *American Machinist*, **100** (6), 160–72.
18. Kibbey, D. R. & Moore, H. D. (1960). Ceramic or carbide tools—which? *Am. Mach./Metalworking Mfg.*, **104** (7), 125–6.

Chapter 5

WORKPIECE MATERIALS

5.1. INTRODUCTION

One of the most exciting challenges to metallurgists over the last century has been to increase the strength of materials with minimal increase in cost. This has to some extent been achieved by various means including cold-working, alloying, the use of phase transformations, and the refinement of grain size and microstructure. Many of the principles established for improving materials generally have been applied to the manufacture of improved cutting and forming tool materials. The ideal cutting tool material, as discussed in Chapter 4, should have high strength (and hardness) together with high fracture toughness. This combination of properties is impossible to achieve in practice since, in general, high strength and low fracture toughness, and the converse, are almost synonymous. However, for materials which are machined, from the machining viewpoint generally, low strength and low fracture toughness are required to give low cutting forces, low cutting temperatures and hence low tool wear rates at high rates of production. This chapter explains the main microstructural and material properties of a wide range of metals and alloys and the effect of these properties on machining behaviour.

Machining is one of a number of processes by which metals and alloys are formed and shaped. The properties of materials being machined or formed are primarily influenced by the temperatures and strain rates resulting from the process. It is useful, therefore, to compare the metal cutting process with other forming and shaping processes such as rolling and forging (cold, warm and hot), wire drawing, and extrusion. As a result of work by Ashby & Frost (1) such processes can be summarised by the presentation of a deformation mechanism map which shows the range of temperature and strain rate over which the processes are performed and the stress required for the deformations involved. The deformation map for copper is shown in Fig. 5.1 and similar maps exist

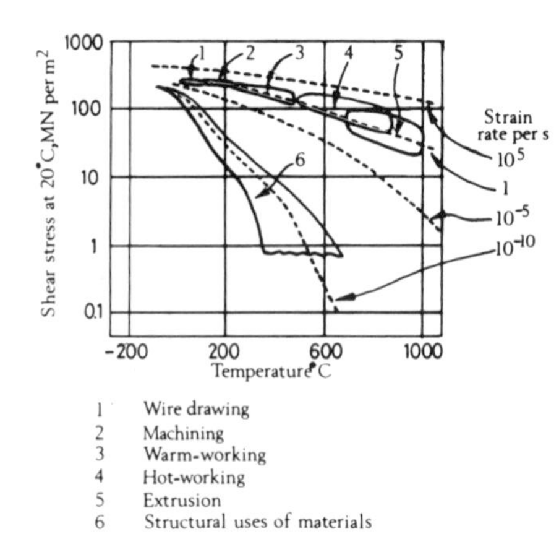

FIG. 5.1. Regions of stress–strain rate and temperature of metal shaping operations for pure copper having a grain size of 0·1 mm (after Ashby & Frost (1)).

for other materials. The map shows the effect of a change in working conditions on the deformation process. In changing from hot- to cold-working at the same strain rate, the shear strength of the material will increase by a factor of four which will result in the forming forces for constant friction increasing by a factor of four. From the diagram it can be seen that increasing the forming rate has little effect on the forces for cold-working but can have a significant effect for hot-working. It can further be seen, for example, that the forces required can be as much as doubled if the strain rate is increased by a factor of 100. During machining at high speeds, much of the heat generated during primary and secondary deformation is retained by the chip. As a result of this the average temperature of the chip can rise to 0·4 of the maximum temperature generated whilst the average chip–tool interface temperature can be much higher than this (0·8 of the maximum temperature generated). During machining, the increase in temperature in both the primary and secondary deformation zones causes a reduction in the shear flow stress of both the parent and chip material but to achieve the higher temperatures, higher strain rates are required which cause an increase in the shear flow stress. As a consequence of these opposing effects, the net result on cutting forces of varying the cutting speed (and hence strain rate) is not significant over the limited range of machining speeds which are practicable. The properties of the workpiece materials which give low tool wear and hence good machinability include the following:

1. *Low yield strength and low work hardening rate*: A workpiece having a low strength and low work hardening rate gives low cutting forces, low contact pressure at the chip–tool interface with a consequential low interfacial temperature.
2. *High thermal conductivity*: A high thermal conductivity assists heat flow away from the secondary deformation zone which also results in a lower interfacial temperature.
3. *Low chemical reactivity with the tool or atmosphere*: The workpiece should not react chemically with the tool or the atmosphere since this can cause tool wear by either low strength tool material alloys being formed or abrasive workpiece alloy particles being formed.
4. *Low fracture toughness*: A low fracture toughness contributes to easy crack formation which results in a reduction in the difficulty of producing broken chips; long continuous chips are undesirable since they present problems of entanglement and disposal. Generally, for a group of similar materials machinability improves as the fracture toughness of the workpiece reduces but there are exceptions to this trend. For example, spheroidal graphite cast irons having higher fracture toughness ($K_{1c} = 45\ \mathrm{MN\,m^{-3/2}}$)* than similar flake graphite cast irons ($K_{1c} = 20\ \mathrm{MN\,m^{-3/2}}$)* give lower cutting tool wear rates.
5. *Absence of unstable built-up-edge during machining*: Unstable built-up-edge formation causes poor surface finish, poorer dimensional accuracy of the machined component and often, due to its hard abrasive nature, leads to higher tool wear rates.
6. *A satisfactory microstructure*: Control of the workpiece microstructure is particularly important since it is one property which can, to some extent, be influenced by the material manufacturer. Clearly, a

*G. N. Gilbert (1982). Fracture toughness data available on cast iron—an assessment, *BCIRA Journal*, 209–17.

material is chosen primarily for its function and the machining behaviour of a material is generally a secondary factor in its selection. However, where a slight degradation of material properties is not important it is possible to select free-machining grades of many metals and alloys. These free-machining grades usually have deliberate additions of sulphides and/or lead. Hard abrasive particles such as carbides and nitrides are undesirable since they are responsible for abrasive tool wear. Hard angular particles are more abrasive than hard rounded particles which indent rather than machine the surface on a microscale.

The efficiency with which metals and alloys are machined can be measured effectively by assessing the power required to machine a unit volume of material in unit time (specific power consumption). A collection of data for a wide range of important engineering materials is shown in Fig. 5.2 (2). Apart from indicating the relative ease of machining steels, cast irons, and magnesium alloys, for the benefit of the user, this information is of use to the machine tool designer in choosing the appropriate power for particular applications of the machine tool.

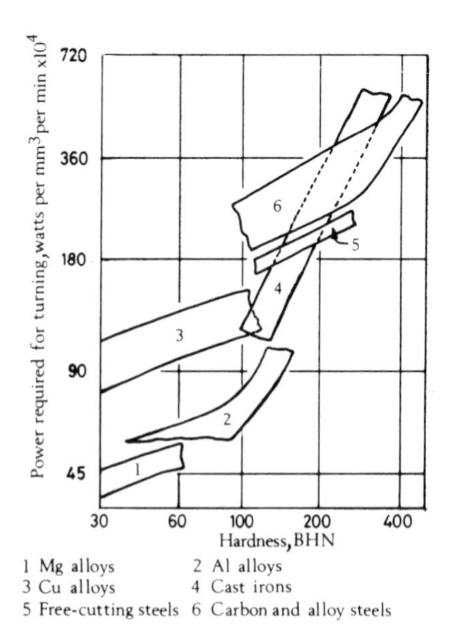

FIG. 5.2. Effect of hardness on the power required to machine different materials.

5.2. FERROUS MATERIALS

5.2.1. Carbon Steels

5.2.1.1. Composition, Microstructure and Mechanical Properties

The mechanical properties of plain carbon steels depend on their composition and microstructure. In annealed steels the amount of pearlite increases with increasing carbon content which causes an increase in yield strength. The influence of increasing carbon content of hot-rolled and cold-drawn carbon steels on the power requirement for machining and on the shear strength is shown in Fig. 5.3 where it can be seen that

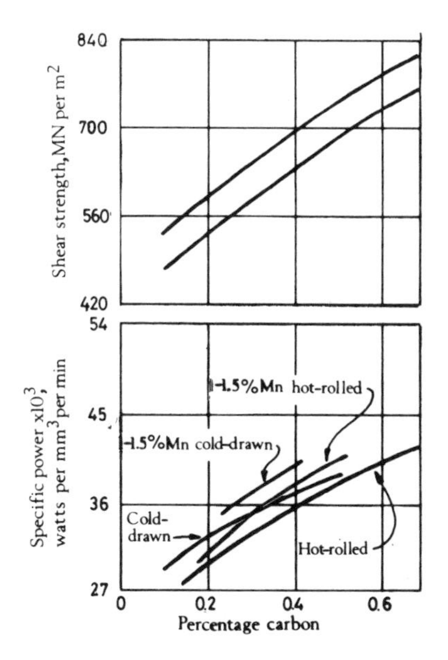

FIG. 5.3. Comparison of shear strength and specific power as a function of carbon content (after Merchant (3)).

there is an approximately linear relationship between the specific power required for machining and the shear strength of the workpiece which, in turn, is nearly linearly related to the carbon content of the steel (3). Plain carbon steels with a carbon content below 0·15% are very soft and adhere strongly to the cutting tool. The machining behaviour can be improved by cold-drawing which increases the strength and reduces the ductility of the workpiece. Medium carbon steels machine best in the coarse pearlite or spheroidised carbide conditions where the coarse structure gives lower strength. Steels with a carbon content in excess of 0·6% machine best in the fully spheroidised condition. As a general rule, tool wear rates increase in a consistent manner as the carbon content of the workpiece is increased beyond 0·35%. To minimise tool wear, for a given rate of production, it is important that steels are chosen which have the minimum carbon content consistent with the strength requirement of the finished component. The surface finish depends not only on the carbon content but also on the cutting operation, the tool geometry and the cutting conditions. When slower cutting processes such as milling and shaping are used, the surface finish is found to be virtually independent of carbon content. Conversely, for example, for high speed turning operations, surface finish improves with increasing carbon content up to 0·35% but deteriorates as the carbon content is further increased.

For machining purposes, the *Metals handbook of machining* of the American Society for Metals (2) classifies steels into the following broad categories:

Low carbon steels
Medium and high carbon steels
Resulphurised low carbon steels
Resulphurised medium carbon steels
Medium and high carbon low alloy steels
Resulphurised low alloy steels
Leaded low alloy steels

The handbook gives details of the nominal speeds and feeds for machining these categories of steels for various hardness levels. The reader is referred to the *Metals handbook of machining* for the very detailed information which is given. As a general rule the recommended cutting speed for both roughing and finishing operations falls with an increase in workpiece hardness. A typical relationship for the effect of hardness on the speed and feed for both rough and finish turning is shown in Fig. 5.4. The combinations of speed and feed are for a tool life of 60 min using a cemented carbide cutting tool (79% WC, 15% TiC, 6% Co).

5.2.2. Free-Machining Steels

Free-machining steels are produced with additions of 0·1 to 0·3% sulphur or 0·1 to 0·35% lead or combinations of both to reduce cutting forces,

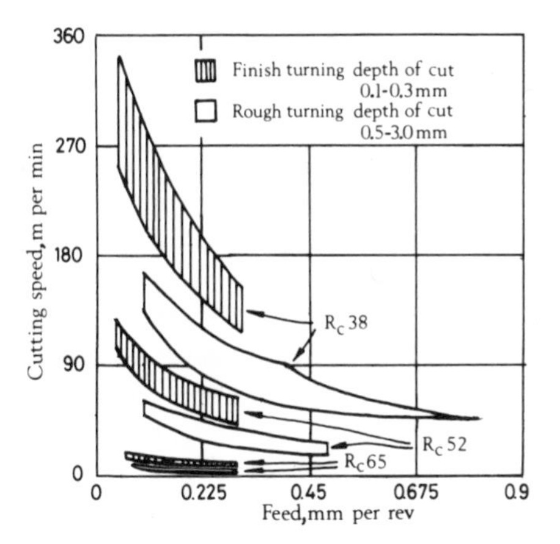

FIG. 5.4. Effect of hardness on speeds and feeds in rough and finish turning (2).

cutting temperatures and hence tool wear rates. These additions allow, for the same wear rate, higher cutting speeds than those used for the equivalent non-free-machining grade. They may also result in the elimination of secondary machining processes because of improved surface finish and this is particularly true for free-machining plain carbon steels.

5.2.2.1. The Influence of Manganese Sulphide Inclusions

Sulphur in free-machining steels is present as manganese sulphide inclusions. Keissling (4) has summarised the role of inclusions in enhancing machinability as follows:

(1) The inclusions should act as stress raisers in the shear plane to initiate cracking and so embrittle the chip;
(2) The inclusions should participate in the flow of metals in the secondary deformation zone to increase the shear but should not disrupt the plastic flow of the metal and abrade the tool surface;
(3) Inclusions should form a diffusion barrier on the face of the tool at the temperature of the chip–tool interface during cutting; and
(4) Inclusions should not prevent the production of smooth workpiece surfaces nor should they cause abrasion of the flank of the cutting tool.

Apart from additions of sulphur and lead, the machinability of steels is further enhanced by additions of bismuth, selenium, tellurium and phosphorus. The machinability of certain steels can be enhanced by deoxidation using calcium; calcium deoxidation modifies the silicate inclusions resulting from deoxidation and these modified inclusions can then act as an interfacial lubricant and this results in a reduction in tool wear. Inclusions which are detrimental to machinability are alumina (Al_2O_3), silica (SiO_2), nitrides, carbides and carbonitrides since these inclusions are very hard and abrasive to cutting tools.

Keane (5) has shown that the wear rate of high-speed-steel cutting tools when machining low-carbon free-machining steels is strongly dependent on the volume fraction of manganese sulphide inclusions up to a volume fraction of 1% (0·2%S) but that for volume fractions above this level, wear rates do not decrease further although the chip form and surface finish continue to improve (Fig. 5.5). It is accepted that manganese sulphide inclusions act as stress raisers in the primary deformation zone which leads to a reduction in the shear stress in this zone.

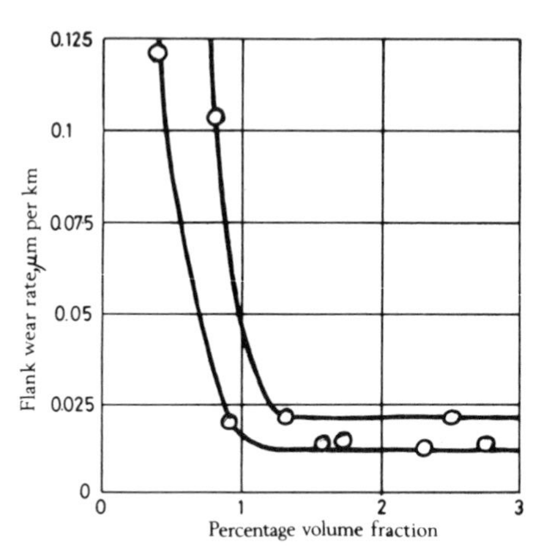

FIG. 5.5. The effect of volume fraction of manganese sulphide inclusions on the wear rate of a high-speed-steel tool (after Keane (5)).

This leads to more efficient machining and a tendency for chip breaking to improve with increasing volume fraction of inclusions. There is much metallographic evidence that manganese sulphide inclusions deposit on the cutting surfaces of the tool and that this leads to a reduction in friction and hence interfacial temperatures. This causes reduced tool wear since, as has previously been pointed out, tool wear rate is very temperature dependent. Further, the manganese sulphide inclusions on the flank and the face of the tool, besides reducing friction, also physically protect these surfaces.

A number of workers (6, 7) have shown that the oxygen content of resulphurised steels (controlled by the degree of deoxidation during steelmaking) has an influence on the nature of the inclusions and on the machinability of the steel. For high oxygen levels the inclusions remain globular during hot-working rather than becoming elongated; this indicates that the inclusions have a low plasticity compared to the matrix during hot-working. The better machinability of steels containing globular manganese sulphide inclusions is due to the absence of abrasive deoxidation products rather than any inherant properties of the inclusions. Apart from improving machinability, manganese sulphide inclusions are otherwise undesirable since they lead to reduced fatigue strength and corrosion resistance.

5.2.2.2. Influence of Lead

Lead is only slightly soluble in steel and appears as a fine dispersion of lead particles either alone or associated with manganese sulphide inclusions. At hot-working and machining temperatures lead is more deformable than manganese sulphide inclusions and in resulphurised leaded free-machining steels appears as long tails on the sulphide inclusions. Lead behaves in a similar manner to manganese sulphide in reducing the shear strength of the material and producing chips which curl tightly and are easy to break. Shaw *et al.* (8) in a study of leaded free-machining steels have shown that more tightly curled chips have a smaller apparent area of contact between chip and tool than the equivalent non-free-machining grade. This could have been implied since an increase in chip curl is accompanied by a reduction in tool face friction as a result of a reduction in apparent area of interfacial contact. The thinner chips formed as a result of decreased friction give lower forces between chip and tool and hence reduced temperatures and wear rates.

Free-machining additives act as internal lubricants during cutting. Metallographic studies have shown that these additives, unlike externally applied lubricants, present themselves during the cutting process at the right place, i.e. at the chip–tool interface. For this reason, internal lubricants such as lead are effective in high speed cutting whereas, for continuous cutting, externally applied lubricants are virtually ineffective, as lubricants, for all but very low cutting speeds since they require time to 'penetrate' to the chip–tool interface. It has been shown that whilst lead in a uniformly dispersed form has little effect on the room temperature properties of the parent material such as ductility, strength and toughness, it does cause embrittlement at temperatures near to the melting point of lead. An Auger electron spectroscope study of chips of leaded free-machining steels showed that lead one atomic layer in thickness was present on the surface of the chips which indicates that reduced tool wear can be effected by layers of only atomic dimensions (9).

In summary, the beneficial effects of lead are due to a combination of reduced shear strength of the workpiece, reduced friction between chip and tool due to monolayer lubrication and physical protection of the potential wear surfaces.

The adverse effects of free-machining additives on the strength and fatigue properties of materials have been summarised by Joseph & Tipnis (10).

5.2.2.3. Influence of Calcium Deoxidation

During recent years, attention has been given to the development of structural steels having improved machinability and, to this end, it was found by Wicker (11) in Germany in 1967 that steels deoxidised using calcium exhibit improved machinability when cut by a variety of cutting processes.

Konig (12) and Opitz & Konig (13) showed that cutting tool wear was greatly influenced by the method of deoxidation of the steel. Steel to AISI 1045 specification, a 0·45% plain carbon steel, which had been deoxidised with calcium was found to give lower cemented carbide cutting tool wear rates than an equivalent silicon-deoxidised steel. The improved machining behaviour of calcium-deoxidised steels is due to the formation of wear retarding surface layers which develop at the chip–tool interface. The effect of cutting speed on the life of a P10 cemented carbide cutting tool measured by criteria based on the development of a crater depth of 0·03 mm and a flank wear land of 0·2 mm during a turning operation is shown in Fig. 5.6 (14) where it can be seen that for a given cutting speed, tool life was extended two to three times when machining the calcium-deoxidised steels. The results of Fig. 5.6 were obtained by the Japanese Working Group on Machinability under the chairmanship of T. Sata (14). This study involved both calcium- and silicon-deoxidised steels which had been produced in Japan and a study of a German equivalent steel which had been calcium deoxidised. A chemical analysis of these three steels is given in Table 5.1.

The Japanese working group showed that for calcium-deoxidised steels the improved tool life extended over a much wider speed range (75–300 m min^{-1}) and to higher cutting speeds than those for which resulphurised and leaded steels are effective. They found that an improved tool performance of P10, P20, M20 and titanium carbide-based

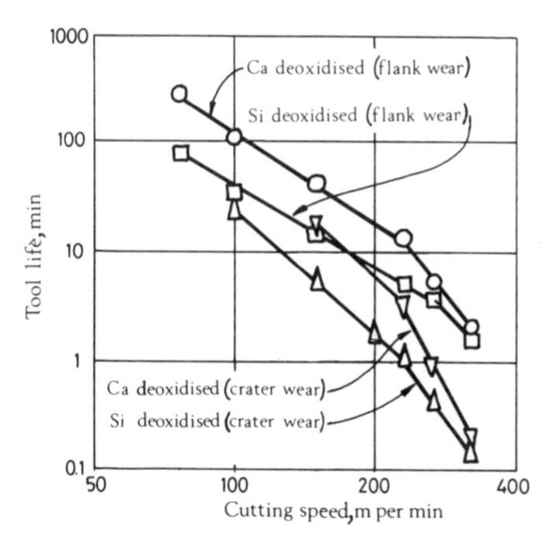

FIG. 5.6. Cutting speed–tool life relationship for a 0·4% carbon steel deoxidised by silicon and calcium (after Sata *et al.* (14)).

cemented carbide cutting tools was obtained when machining the calcium-deoxidised steels although no improvement was observed for high-speed-steel cutting tools. In contrast, it has been reported that small additions of lead and sulphur, either alone or together, further improve the machinability of calcium-deoxidised steels for both cemented carbides and high-speed-steel cutting tools (15).

The mechanism of improved machinability was studied in detail by the Japanese group (14) and this included a chemical analysis of the inclusions and a study of the basic cutting mechanism. They found that the surface finish, chip–tool contact length and shear angle remained constant for similar cutting conditions for the calcium- and silicon-deoxidised steels. Using electron diffraction analytical techniques they examined the non-metallic inclusions in the calcium-deoxidised steels of both Japanese and German manufacture. The chemical analysis of the inclusions is given in Table 5.2.

The differences in chemical analysis of the inclusions found in the German and Japanese calcium-deoxidised steels arise from the significantly higher calcium content of the Japanese steel. Importantly, both Konig (12) and Sata *et al.* (14) found that the cutting surfaces of the tools were covered with a protective layer which resulted from the non-metallic inclusions present in the workpiece. The Japanese workers found that this layer contained calcium, aluminium and manganese when machining the Japanese steels and additionally contained silicon when machin-

TABLE 5.1
CHEMICAL ANALYSIS OF CALCIUM- AND SILICON-DEOXIDISED STEELS (14)

Steel	*Chemical composition (wt %)*										
	C	*Si*	*Mn*	*P*	*S*	*Cu*	*Ni*	*Cr*	*Ca*	*Al*	*O* (ppm)
Japanese calcium-deoxidised	0·47	0·33	0·73	0·014	0·012	0·15	0·06	0·11	0·0047	0·007	86
Japanese silicon-deoxidised	0·47	0·30	0·69	0·017	0·014	0·17	0·06	0·10	0·0002	0·005	86
German calcium-deoxidised	0·45	0·35	0·69	0·027	0·033	0·17	0·06	0·06	0·0012	0·004	53

TABLE 5.2

CHEMICAL ANALYSIS OF NON-METALLIC INCLUSIONS IN JAPANESE- AND GERMAN-MANUFACTURED CALCIUM-DEOXIDISED STEELS (14)

German steel		*Japanese steel*	
Compound	*wt%*	*Compound*	*wt%*
$3\,CaO\;2\,SiO_2$	77	$2\,CaO\;SiO_2$ (γ)	44
$2\,CaO\;SiO_2$	9	$2\,CaO\;SiO_2$ (β)	6
$2\,CaO\;Al_2O_3\;SiO_2$	9	$CaO\;SiO_2$ (α)	14
Unidentified	5	$5\,CaO\;3\,Al_2O_3$	3
	100	$Al_2O_3\;SiO_2$	6
		SiC	6
		Unidentified	21
			100

ing the German steel. Opitz & Konig (13) analysed the surface layer which developed when machining both conventionally silicon-deoxidised and calcium-deoxidised steels. The chemical analysis of these layers is given in Table 5.3.

Opitz & Konig (13) found that calcium, silicon, aluminium and manganese occurred as CaO, SiO_2, Al_2O_3 and MnO combined in solid solutions of these oxides. Making reasonable assumptions on the basis of the chemical analysis of the layer, they estimated that the layer could be considered to be a binary solid solution of the $CaO–SiO_2$ system containing dissolved Al_2O_3 which had a melting point of approximately 1500°C. From their metal cutting studies, they found that built-up-edge formation did not occur at cutting speeds higher than approximately 70 m min^{-1} which would allow the layer to form at cutting speeds greater than this. Using metallographic and electron optical techniques

TABLE 5.3

CHEMICAL ANALYSIS OF THE LAYERS DEVELOPED ON CEMENTED CARBIDE CUTTING TOOLS WHEN MACHINING SILICON- AND CALCIUM-DEOXIDISED STEELS (13)

	Chemical analysis of layer (wt%)						
Workpiece	*Ca*	*Al*	*Si*	*Mn*	*S*	*Fe*	*Ti*
Calcium-deoxidised	24–28	10–16	3–4	0·5–4	0·5–20	0·5–20	1
Silicon-deoxidised	5–16	5–11	11–23·5	3–24	0·5–1·5	0·1–6	Trace

they detected the layer at cutting speeds greater than 100 m min^{-1}. It was considered that the oxide inclusions present in the workpiece under the high temperatures and pressures generated in high speed machining react with the cemented carbide cutting tool to give a strongly adherent film which grows and is replenished under the dynamic conditions of machining and affords protection by acting as both a diffusion and a physical barrier to the wear processes.

Calcium-deoxidised steels containing 0·3 to 0·55% carbon are widely used in Japan as a substitute for the leaded grades which are restricted in use. The calcium-deoxidised steels are produced in both plain carbon and low alloy carburising grades as coarse-grained steels. The heat treatment properties, response to carburising, hardenability and mechanical properties, apart from impact strength, are equivalent to conventionally deoxidised steels. The main advantages of the calcium-deoxidised steels over resulphurised and leaded grades are their better conventional fatigue and contact fatigue properties. The development of fine-grained calcium–aluminium-deoxidised steels in the USA (trade name Cal De Ox) for automotive gears has been described by Tipnis *et al.* (16). These steels were developed to avoid large entrapped aluminium oxide inclusions, particularly in continuously cast steels; these inclusions caused very high tool wear rates on high-speed-steel gear cutters. The improved machining behaviour of Cal De Ox steels is due to the presence of calcium aluminate inclusions which are softer and, therefore, less abrasive than alumina inclusions. It has also been reported (15) that the improved machining performance of Cal De Ox steels is due in part to the envelopment of the calcium aluminate inclusions by a soft (MnCa)S phase. This contrasts with conventionally deoxidised aluminium–silicon steels where separate hard aluminium and eutectic sulphide inclusions are present.

The mechanical properties of calcium-deoxidised steels are not significantly different to those of conventionally deoxidised equivalent steels; this is clearly an advantage since calcium-deoxidised steels exhibit better machinability. Unlike steels containing additions of sulphur and lead the calcium-deoxidised steels, when machined, do not have an improved chip form which would lead to better chip forming, breaking and removal. Ito *et al.* (17) have shown that small additions of sulphur (up to 0·075%) and/or lead (up to 0·06%) to calcium-deoxidised steels have a profound beneficial effect on the chip form. Further, sulphur and lead appear to act independently and their effects on improving the chip form are additive.

5.2.3. Stainless Steels

Stainless steels can be broadly divided into four groups, the austenitic, ferritic, martensitic and precipitation-hardening types. Detailed recommended machining conditions for all the main types of steels within these groups are given in the *Metals handbook of machining* (2) and are also commented on by Tipnis (18). The important properties which influence the machining behaviour of stainless steels include the following:

(1) Stainless steels have a higher tensile strength and a greater spread between yield and fracture strength than plain carbon steels. The energy required, therefore, to machine stainless steels is higher than for non-hardened plain carbon steels.
(2) Austenitic stainless steels have high work-hardening rates and low thermal conductivity and the high work-hardening rate causes higher energy consumption compared with non-hardened plain carbon steels. The low thermal conductivity causes higher temperature gradients within the chip and the higher temperature generated in the secondary deformation zone causes higher chip–tool interfacial temperatures which result in increasing diffusion wear rates (diffusion wear is very strongly temperature dependent).
(3) The higher alloy stainless steels containing abrasive carbides cause accelerated tool wear. Annealed martensitic stainless steels (such as AISI 1410) machine similarly to low alloy carbon steels but any primary carbides present in these grades cause excessive tool wear. For finish machining of martensitic steels in the hardened and tempered condition, high cutting speeds should be avoided since this could cause overtempering. The ferritic grades such as AISI 430 machine in a similar way to the annealed martensitic grades. The austenitic grades such as the AISI 300 series have a lower machinability rating than the martensitic and ferritic grades although the free-machining properties of stainless steels are achieved by the use of free-machining additives.

5.2.3.1. Effect of Microstructure on Machinability

The presence of hard particles causes rapid tool wear by a mechanism of abrasion. Soft ductile inclusions reduce the energy consumed during cutting by lowering the ductility of the workpiece by an action of internal lubrication which reduces cutting temperatures and hence reduces cutting tool wear rates. The influence of hard carbides and soft non-metallic inclusions on the machining behaviour of three austenitic stainless steels has been examined by Akhtar & Mills (19). The tool life cutting speed curves for a plain AISI 302 steel, a plain AISI 302 steel containing excess alloy carbides and a resulphurised AISI 302 steel are shown in Fig. 5.7. The lowest tool wear rates are found when machining the resulphurised steel whilst the non-resulphurised steel containing alloy carbides gave the highest cutting tool wear rates.

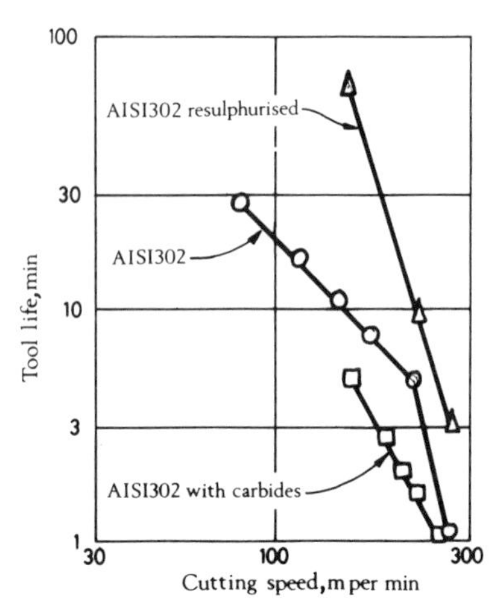

FIG. 5.7. Cutting speed–tool life relationship for three austenitic stainless steels.

The effect of ductility on machinability has been reviewed by Boulger (20) and discussed by Tipnis (18). Tipnis suggests that for stainless steels a lower ductility improves machinability by causing primary deformation zone fracture which makes chip removal easier. He further suggests that the beneficial effects of sulphur, selenium and tellurium which have all been added to stainless steels to improve machinability do this by causing a reduction in ductility.

The composition, quantity and physical properties of non-metallic inclusions influence the machining properties of stainless steels. The presence of appreciable amounts of chromium and to a lesser extent molybdenum allow the formation of complex sulphide non-metallic inclusions. These complex sulphides have different properties to the manganese sulphide inclusions found in low carbon free-machining steels. Garvin & Larrimore (21) studied the effect of manganese content on the machining properties of AISI 435 stainless steels and found that steels containing manganese-rich sulphides gave longer tool lives when being machined using a T1 high-speed-steel tool. Studies by Kovach & Moskowitz (22,23) on austenitic (AISI 303) and martensitic (AISI 416) stainless steels have shown that the chemical composition of the non-metallic inclusions has a marked effect on the machinability of these materials. The studies on AISI 303 stainless steel (22) showed that cutting tool wear measured in drilling tests decreased with increasing manganese content of the inclusions. A high manganese to sulphur ratio was considered to give softer inclusions and for a significant improvement in machining behaviour the manganese content should be greater than 2%. For martensitic stainless steels Kovach & Moskowitz (23) showed that a high manganese content gave reduced tool wear for both drilling and turning operations. With increasing manganese content, the composition of the inclusions changes from CrS to (Fe Mn) Cr_2S_4 to (Mn Fe Cr)S to MnS. The fall in hardness of the inclusions with an increased manganese content causes increased plasticity of the inclusions and the effect of the Mn to S ratio on the machining behaviour of the martensitic AISI 416 stainless steel is shown in Fig. 5.8 (23). Kovach & Moskowitz concluded

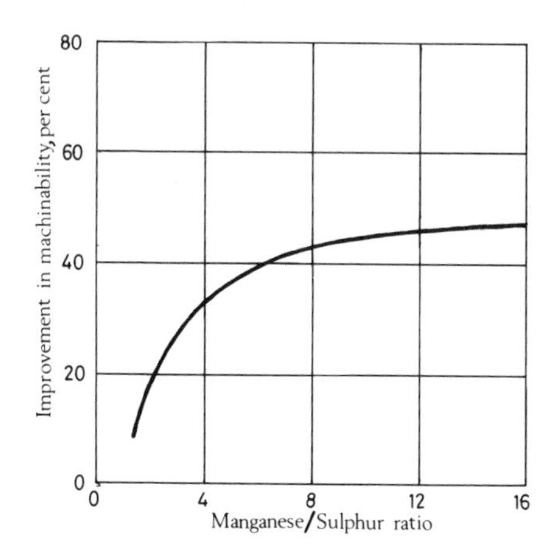

FIG. 5.8. The effect of the manganese to sulphur ratio on the machinability of AISI 416 stainless steel (after Kovach & Moskowitz (23)).

that a manganese to sulphur ratio of greater than seven is required to obtain the maximum improvements in machinability.

Metallographic studies have shown that manganese sulphide inclusions are deformed into thin plates whereas chromium-rich inclusions break up during deformation into discrete particles. Thus it has been suggested that whereas manganese sulphide inclusions wrap themselves around the tool face and flank and offer protection, most other types of inclusion tend to travel up the face and across the flank and abrade them.

The effect of selenium (24), sulphur and selenium (25) and sulphur, selenium and tellurium (26) on the machinability of stainless steels has also been studied by a number of workers. The results of Belov *et al.* (26) are given in Fig. 5.9 where the V_{60} cutting speed is plotted against the

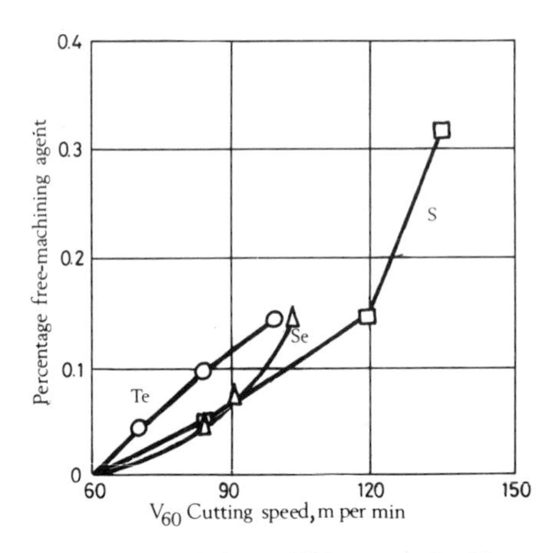

FIG. 5.9. The effect of free-machining additives on the V_{60} cutting speed of an austenitic stainless steel (after Belov (26)).

level of sulphur, selenium and tellurium. From these results it can be concluded that sulphur is the most effective, and tellurium the least effective, addition. However, it is worth noting that the selenium-bearing steel gave the best surface finish of the three free-machining additives investigated. Tellurium is found in the elemental form in steels since it is only slightly soluble in manganese sulphide; selenium forms manganese selenide which is soluble in manganese sulphide. Both tellurium and selenium impair the hot-working properties of the steel and must be added under carefully controlled melting conditions since their oxides are highly toxic. Although lead is not added extensively to stainless steels since it has a strong tendency to segregate it has been shown by Clark (25) that leaded AISI 303 steel, when machined, has a better surface finish than an equivalent resulphurised grade.

5.2.4. Cast Irons

5.2.4.1. Effect of Composition and Microstructure

Cast irons show a wide range of machining behaviour which depends upon composition and microstructure. The white cast irons containing large quantities of hard carbides are difficult to machine and are responsible for high tool wear rates whilst the ferritic grey cast irons containing primarily iron and graphite are very easy to machine. For grey cast irons the effect of matrix microstructure on the 'tool life index' has been determined from United States Air Force Machinability Reports and is given in the *Metals handbook of machining* (2). Table 5.4 lists the

TABLE 5.4
THE EFFECT OF MATRIX MICROSTRUCTURE AND HARDNESS ON THE CUTTING TOOL LIFE INDEX OF GREY CAST IRONS (2)

Matrix microstructure	*Brinell hardness*	*Tool life index*
Ferrite	120	20
50% Ferrite 50% pearlite	150	10
Coarse pearlite	195	2
Medium pearlite	215	1·5
Fine pearlite	218	1
Fine pearlite with 5% excess iron carbide	240	0·3

tool life index for different matrix microstructures of grey cast irons containing 2·5% of graphite and of different hardness values. It can be seen from the table that tool wear rates increase as the hardness of the matrix increases. The hardness of the matrix increases with increasing proportions of pearlite in pearlite/ferritic matrices and with decreasing pearlite interlamellar spacing; a further increase in tool wear rates occurs when the pearlitic matrix contains excess iron carbide. Graphite in cast iron reduces ductility and this causes chip breakage; this is the main reason for the better machining behaviour of grey cast irons compared with steels having equivalent matrix microstructures. This better machining behaviour of grey cast irons is illustrated in Fig. 5.10 which is a machining chart for a tungsten carbide–6% Co cemented carbide machining a pearlitic grey flake cast iron. Comparing Fig. 5.10 with Fig. 4.7—the machining chart for a 0·4% plain carbon steel—it can be seen that rapid cratering occurs at higher combinations of speed and feed for

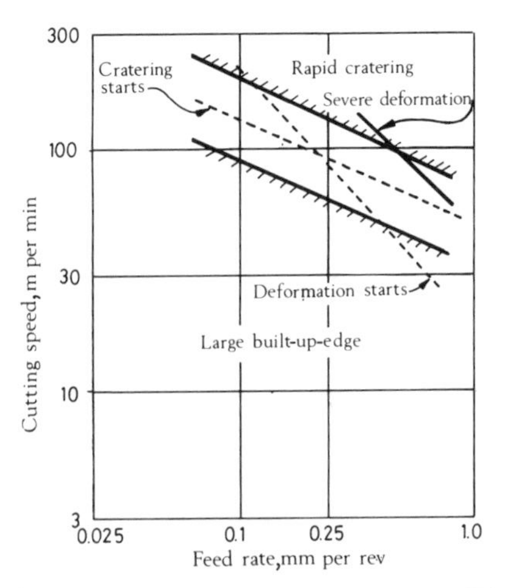

FIG. 5.10. Machining chart for a tungsten carbide plus 6% cobalt cutting tool turning pearlitic grey cast iron (after Trent (27)).

the grey cast iron. Trent (27) showed that when cutting with both high-speed steels and cemented carbides built-up-edges form for irons which persist to higher cutting speeds than for steels (compare Fig. 5.10 with Fig. 4.17)

Nodular cast irons have higher tensile strength than equivalent flake graphite cast irons although for turning operations with cemented carbide cutting tools it has been shown that they give lower tool wear rates. Figure 5.11 shows the volume of metal removed in the development of 0·75 mm flank wear land when machining flake and nodular cast irons of equivalent hardness. These results show that nodular cast irons give lower tool wear for all conditions except at the higher cutting speeds for the harder irons.

The machining behaviour of cast irons is influenced by composition largely through the effects of alloying elements in promoting graphite or carbide formation.

Phosphorus, however, which is added to promote fluidity in casting, is, in quantities greater than 0·15%, considered to be undesirable from the machinability viewpoint since it forms the hard iron–iron phosphide eutectic. The effect of this on the wear rate of cutting tools can be reduced by transforming the pearlitic matrix surrounding the eutectic to a ferritic one by annealing. This soft material allows the hard eutectic particles to embed in the chip during cutting rather than them abrading the tool.

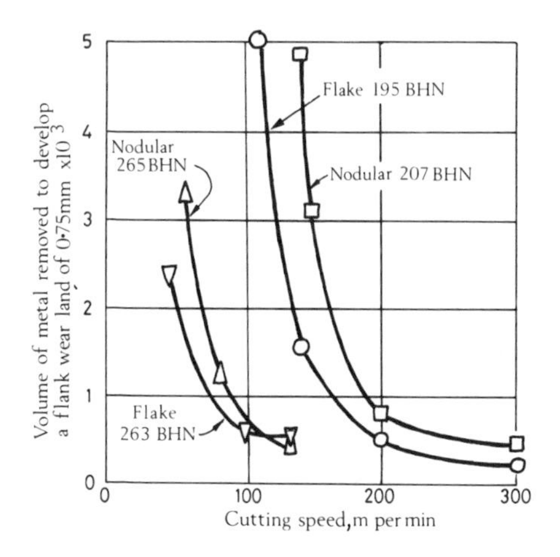

FIG. 5.11. Machinability of flake and nodular cast irons (2).

5.3. TITANIUM ALLOYS

Titanium alloys are classed as difficult-to-machine alloys and the properties listed below contribute to their poor machining behaviour:

(1) Titanium has a very low thermal conductivity (one-sixth that of steel) which results in high temperature gradients at the chip–tool interface.
(2) The chip–tool contact area when machining titanium is small compared with that for steel. This also increases both the temperature and the temperature gradient at the chip–tool interface. Further, the small contact area compared with the relatively high strength of these alloys results in very high contact pressures.
(3) Titanium and its alloys have a tendency to pressure weld to cutting tools and this contributes to cutting tool wear. Titanium reacts chemically with oxygen and nitrogen at elevated temperatures and this chemical reactivity contributes to galling and abrasive wear of the cutting tool and to the pyrophoric behaviour of small particles of titanium.
(4) The relatively low modulus of elasticity of titanium and its alloys can lead to distortion during machining.

For these reasons it is necessary to machine titanium and its alloys at lower cutting speeds than those which would be appropriate for steels having similar hardness levels. Titanium alloys are normally classified into three groups designated α, $\alpha+\beta$, and β alloys. Zlatin & Field (28) have compared the machining behaviour of typical alloys from each of these groups with that of a medium carbon content low alloy steel, AISI 4340, having a hardness of 300 BHN. Cemented carbide tools were used for turning and face milling operations and high-speed-steel tools were used for drilling. The machining time to machine unit volume of workpiece compared with the low alloy steel is given in Table 5.5 for the three different operations. From this table it can be seen that α-phase alloys can take up to 2·5 times the machining time of the AISI 4340 steel whilst the β-phase alloys can require up to 10 times the machining time.

TABLE 5.5

MACHINING TIMES[a] TO MACHINE UNIT VOLUME OF WORKPIECE FOR VARIOUS MACHINING OPERATIONS AND VARIOUS TITANIUM ALLOYS (28)

Alloy	*Hardness*	*Turning*	*Face milling*	*Drilling*
Commercial purity titanium	175	0·7	1·4	0·7
α-alloy Ti, 8% Al, 1% Mo, 1% V	300	1·1	2·5	1
$\alpha+\beta$-alloy Ti, 6% Al, 4% V	350	2·5	3·3	1·7
β-alloy Ti, 3% Al, 11% Cr, 13% V	400	5	10	10

[a]Values of machining time are relative to the time to machine an AISI 4340 low alloy steel for the appropriate machining process.

5.4. NICKEL-BASED ALLOYS

Nickel-based alloys, particularly the creep-resistant ones, are amongst the most difficult alloys to machine. The main reason for the poor machinability of these alloys is the high work-hardening rate and the presence of hard abrasive phases such as titanium carbide, niobium carbide and the Ni_3AlTi phase. The effect of percentage cold reduction on the hardness of several nickel-based alloys, mild steel, copper and aluminium is shown in Fig. 5.12. The nickel-based alloys also retain their strength at elevated temperatures and this results in high cutting forces even at high cutting speeds for which high temperatures are generated. The effect of temperature on the strength of a range of nickel-based alloys and copper is shown in Fig. 5.13.

Because nickel-based alloys are difficult to machine, it is usual to employ low cutting speeds to reduce cutting tool wear rates. After a

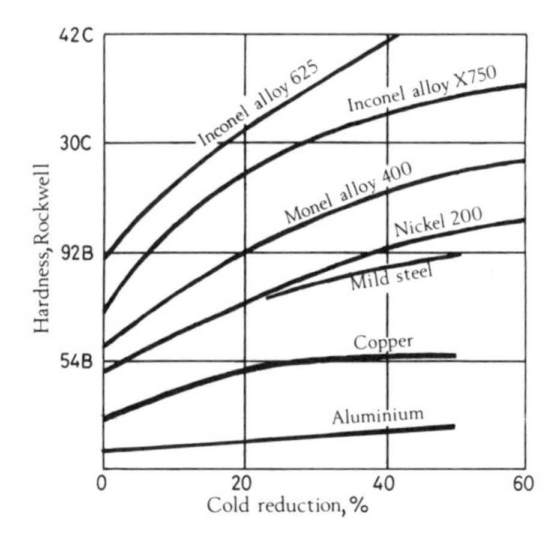

FIG. 5.12. Effect of cold reduction on the hardness of various nickel-based alloys, mild steel, copper and aluminium.

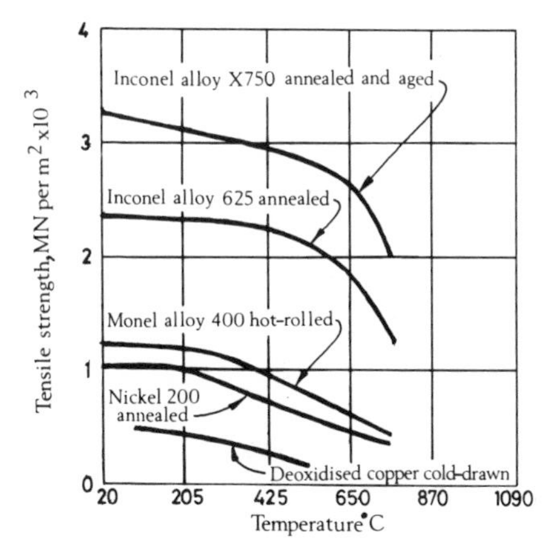

FIG. 5.13. Effect on tensile strength of temperature for various nickel-based alloys and copper.

machining pass the newly generated workpiece surface is harder, due to work hardening, than the underlying material. For this reason the choice of feed is very important; if the feed rate is too low then the tool is effectively cutting continuously through work-hardened material which was generated on the previous cut. Conversely, at very high feed rates, even if the surface finish were acceptable, the forces acting on the cutting tool could be too high and cause catastrophic tool failure. A compromise

TABLE 5.6

CLASSIFICATION OF NICKEL ALLOYS BASED ON MACHINING CHARACTERISTICS (2)

Alloy	*Ni*	*C*	*Mn*	*Fe*	*S*	*Si*	*Cu*	*Cr*	*Ti*	*Al*	*Co or Nb*	*Mo or Mg*
Machining Group A												
Nickel 200	99·5	0·06	0·25	0·15	0·005	0·05	0·05	—	—	—	—	—
Nickel 201	99·5	0·01	0·20	0·15	0·005	0·05	0·05	—	—	—	—	—
Nickel 204	95·2	0·06	0·20	0·05	0·005	0·02	0·02	—	—	—	4·50 Co	—
Nickel 205	99·5	0·06	0·20	0·10	0·005	0·05	0·05	—	0·02	—	—	0·04 Mg
Nickel 211	95·0	0·10	4·75	0·05	0·005	0·05	0·03	—	—	—	—	—
Nickel 220	99·5	0·06	0·12	0·05	0·005	0·03	0·03	—	0·02	—	—	0·04 Mg
Nickel 230	99·5	0·09	0·10	0·05	0·005	0·03	0·01	—	0·003	—	—	0·06 Mg
Nickel 233	99·5	0·09	0·18	0·05	0·005	0·03	0·03	—	0·003	—	—	0·07 Mg
Machining Group B												
Monel 400	66·0	0·12	0·90	1·35	0·005	0·15	31·5	—	—	—	—	—
Monel 401	44·5	0·03	1·70	0·20	0·005	0·01	53·0	—	—	—	0·50 Co	—
Monel 402	58·0	0·12	0·90	1·20	0·005	0·10	40·0	—	—	—	—	—
Monel 403	57·5	0·12	1·80	0·50	0·005	0·25	40·0	—	—	—	—	—
Monel 404	55·0	0·06	0·01	0·05	0·005	0·02	44·0	—	—	0·02	—	—
Monel 501, graphitised	65·0	0·23	0·60	1·00	0·005	0·15	29·5	—	0·50	2·80	—	—
Machining Group C												
Nickel 270	99·97	0·02	Trace	Trace	Trace	Trace	Trace	Trace	—	—	—	—
Monel K-500, unaged	65·0	0·15	0·60	1·00	0·005	0·15	29·50	—	0·50	2·80	—	—
Inconel 600	76·0	0·04	0·20	7·20	0·007	0·20	0·10	15·8	—	—	—	—
Inconel 604	74·0	0·04	0·20	7·20	0·007	0·20	0·10	15·8	—	—	2·0 Nb	—
Incoloy 800	32·0	0·04	0·75	46·0	0·007	0·35	0·30	20·5	—	—	—	—
Incoloy 801	32·0	0·04	0·75	44·5	0·007	0·35	0·15	20·5	1·00	—	—	—
Incoloy 804	42·6	0·06	0·85	25·4	0·007	0·50	0·40	29·3	0·40	0·25	—	—
Incoloy 825	41·8	0·03	0·65	30·0	0·007	0·35	1·80	21·5	0·90	0·15	—	3·00 Mo

Machining Group D-1												
Permanickel 300, unaged	98·6	0·25	0·10	0·10	0·005	0·06	0·02	—	0·50	—	—	0·35 Mg
Duranickel 301, unaged	94·0	0·15	0·25	0·15	0·005	0·55	0·05	—	0·50	4·50	—	—
Ni-span-C 902, unaged	42·0	0·02	0·40	48·50	0·008	0·50	0·05	5·4	2·40	0·65	—	—
Machining Group D-2												
Permanickel 300, aged	98·6	0·25	0·10	0·10	0·005	0·06	0·02	—	0·50	—	—	0·35 Mg
Duranickel 301, aged	94·0	0·15	0·25	0·15	0·005	0·55	0·05	—	0·50	4·50	—	—
Monel K-500, aged	65·0	0·15	0·60	1·00	0·005	0·15	29·50	—	0·50	2·80	—	—
Monel 501, aged	65·0	0·23	0·60	1·00	0·005	0·15	29·50	—	0·50	2·80	—	—
Inconel 700	46·0	0·12	0·10	0·70	0·007	0·30	0·05	15·0	2·20	3·00	28·5 Co	3·75 Mo
Inconel 702	79·5	0·04	0·05	0·35	0·007	0·20	0·10	15·6	0·70	3·40	—	—
Inconel 718	52·5	0·04	0·20	18·00	0·007	0·20	0·10	19·0	0·80	0·60	5·2 Nb	3·0 Mo
Inconel 721	71·0	0·04	2·25	7·20	0·007	0·12	0·10	16·0	3·00	—	—	—
Inconel 722	75·0	0·04	0·55	6·50	0·007	0·20	0·05	15·0	2·40	0·60	—	—
Inconel X-750	73·0	0·04	0·70	6·75	0·007	0·30	0·05	15·0	2·50	0·80	0·85 Nb	—
Inconel 751	72·5	0·04	0·70	6·75	0·007	0·30	0·05	15·0	2·50	1·20	1·00 Nb	—
Ni-span-C 902, aged	42·0	0·02	0·40	48·50	0·008	0·50	0·05	5·4	2·40	0·65	—	—
Machining Group E												
Monel R-405	66·0	0·18	0·90	1·35	0·050	0·15	31·5	—	—	—	—	—

feed in the range 0·18 to 0·25 mm is usually chosen to allow machining of some non-work-hardened workpiece with acceptable forces on the tool. In order to minimise rubbing effects when machining these alloys the cutting tools should have positive rake. The nickel-based alloys have been classified according to their machining behaviour in the machining handbook (2). The composition of the alloys and the machining group to which they belong are given in Table 5.6. Alloys in Group A are dilute alloys containing more than 95% nickel and these are hardened by cold-working. These alloys give their best machining behaviour in the cold-drawn condition when the surface finish produced is also good. Group B alloys consist of the nickel–copper-based alloys which have higher strength than Group A alloys because of solid solution strengthening. The Group C alloys consist mainly of the nickel–chromium–iron alloys which behave in a similar manner to the austenitic stainless steels. Alloys in Groups B and C are best machined in the cold-drawn or cold-drawn and stress-relieved condition. The age hardening alloys are given in Group D which is subdivided into alloys in the unaged condition (subgroup D-1) and alloys in the aged condition with some in both the aged and unaged condition (subgroup D-2). These alloys have very high strength, particularly in the aged condition. Materials in the solution-treated and quenched (soft) condition machine more easily. They are best rough machined in the solution-treated and unaged condition followed by age hardening and finish machining. The Monel R-405 alloy of Group E is a graphitised nickel–copper alloy having good machinability by virtue of the presence of free graphite in the microstructure.

5.5. ALUMINIUM ALLOYS

Aluminium alloys have low melting points and can be machined relatively easily. For the same family of alloys the power required to machine a unit volume of workpiece is generally proportional to the shear strength and ultimate tensile strength. However, for dissimilar materials this correlation does not hold. It has been shown (2) that a 2017-T4 aluminium–copper alloy (ultimate tensile strength 440 MPa) having similar mechanical properties to a hot-rolled low carbon steel (ultimate tensile strength 450 MPa) machines with much lower cutting forces (Fig. 5.14). The lower forces in the case of the aluminium–copper alloys arise because of the lower frictional forces between chip and tool. For commercially pure aluminium, like many pure metals, the cutting forces are generally much higher than those for alloyed aluminium (in

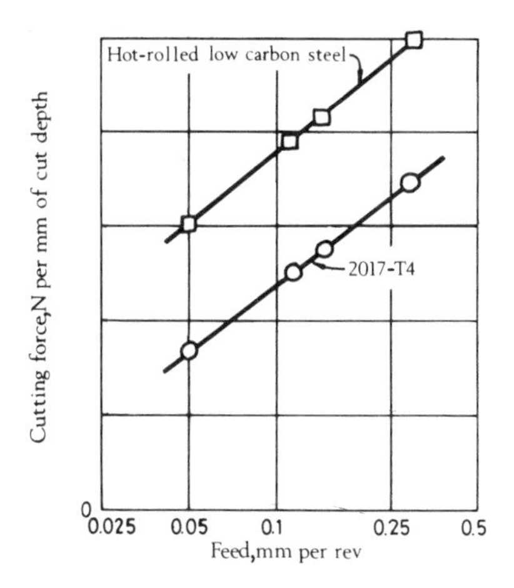

FIG. 5.14. Effect of feed on the cutting force for a low carbon steel and 2017-T4 copper–aluminium alloy.

spite of the lower shear strength of pure aluminium) due to very strong adhesion between the cutting tool and the workpiece.

Aluminium alloys can be broadly divided into the cast and the wrought alloys. Of the cast alloys, alloys with silicon as the main alloying element are the most important group. These alloys contain hard abrasive silicon particles ($\sim 500\,kg\,mm^{-2}$) in the microstructure which can cause tool wear problems; these alloys are more economically machined at low cutting speeds and feeds. The addition of copper to binary aluminium–silicon alloys increases strength but also improves both tool life and chip form by reducing ductility. The aluminium–magnesium and aluminium–zinc–magnesium alloys all have good machinability.

Wrought aluminium alloys are conveniently divided into the non-heat-treatable and the heat-treatable groups; the former are strengthened by working and the latter by precipitation processes. Improved tool lives, but more particularly chip form, when machining these alloys is achieved by additions of the low melting point insoluble metals, tin, bismuth and lead.

5.6. MAGNESIUM AND ITS ALLOYS

Magnesium and its alloys are the easiest to machine. The power required to machine a unit volume of workpiece is an order of magnitude less than that for nickel alloys and is approximately half that required for aluminium alloys. The low power requirements are due to the low shear strength of the alloys and cutting being very efficient—thin segmented chips are produced which present no chip disposal problems. Since fine magnesium chips are liable to ignite at high temperatures (high machining speeds), it is usual to use a cutting fluid. Mineral oil cutting fluids are preferred since water-based coolants promote chip ignition and are, therefore, dangerous.

5.7. COPPER AND ITS ALLOYS

Pure copper is difficult to machine because of the high frictional forces between the chip and the cutting tool (as mentioned before, similar behaviour is observed with pure aluminium and most pure metals). The machining behaviour of copper, like aluminium, is improved by cold-working and alloying. It is appropriate to divide copper alloys into three groups based on their machining characteristics:

(1) Free-machining alloys
(2) Difficult-to-machine alloys
(3) Intermediate-machining alloys

Since copper and copper alloys have only moderate shear strength, machinability is more often based on the type of chip which is produced rather than a tool life criterion. The free-machining copper alloys produce

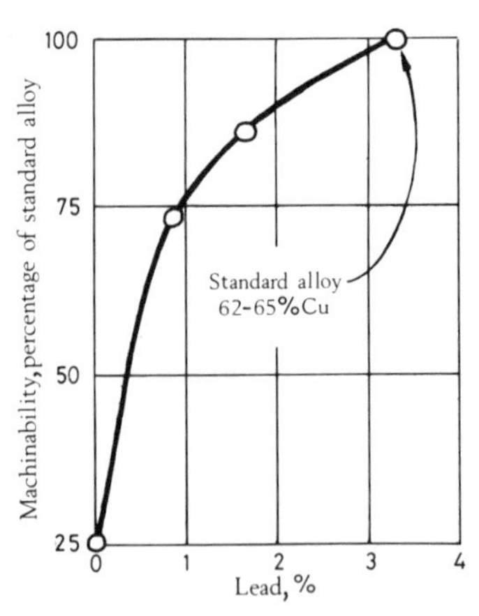

FIG. 5.15. Relationship between lead content and machinability of a brass containing between 62 and 65% copper (2).

short brittle chips which present no difficulty in chip disposal. The difficult-to-machine copper alloys generally produce long continuous chips which tend to snarl and may lead to interruptions of the machining process. The intermediate copper alloys produce open coil or helical chips which can be fragmented using chip forming devices. Alloys belonging to the free-machining group have additions of lead, sulphur and tellurium. In free-machining copper alloys lead is insoluble and appears as globules finely dispersed throughout the matrix. The addition of lead increases the brittleness of the chips and this results in thin fragmented chips. For brass containing between 62 and 65% copper, lead additions up to 3 wt % give a continuous improvement in machining characteristics (Fig. 5.15). A recent Auger electron spectroscope study (29) of copper–40 wt % zinc brasses has shown that the lead additive in quantities greater than 1 wt % gives a complete monolayer of lead on the swarf surface which in the virtual absence of a secondary deformation zone indicates monolayer lubrication by the lead. At bulk lead levels of 2·8 wt %, power consumption is at a minimum and discontinuous chips are produced. For greater concentrations of lead, in addition to the monoatomic lead lubricant layer, a 'thick' layer of lead is produced over 30 to 50% of the swarf surface. Sulphur and tellurium form soft compounds of Cu_2S and Cu_2Te, respectively, which improve chip form by reducing ductility. The effect of sulphur and tellurium on the number of revolutions for a high-speed-steel twist drill to penetrate a distance of 6 mm under a load of 380 N is shown in Fig. 5.16. The

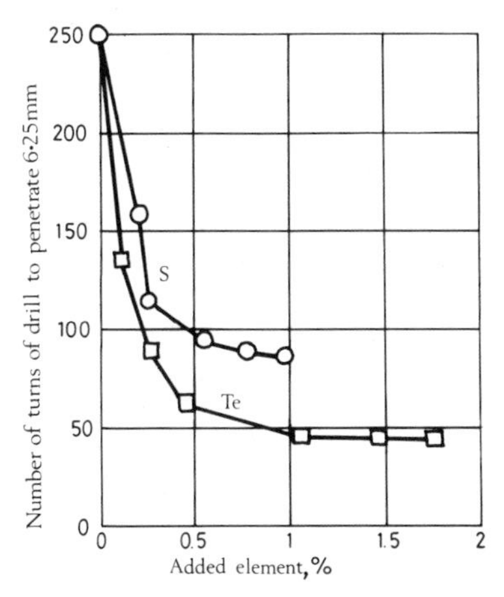

FIG. 5.16. Relationship between the sulphur and tellurium content of copper and the machinability as determined using a constant pressure test (2).

maximum benefit of these additives is obtained with additions of approximately 0·5 wt%. Alloys which have intermediate machining behaviour include the non-leaded copper alloys containing between 60 and 80 wt% of copper. The alloys which are difficult to machine include the low zinc content brasses (single phase), pure copper, the phosphor bronzes and the high strength magnesium and aluminium bronzes which have high iron or nickel contents.

REFERENCES

1. ASHBY, M. F. & FROST, M. J. (1976). *Seven case studies in the use of deformation mechanism maps and the construction of transient maps and structure maps*, Cambridge University Engineering Department, CUED/C, MATS/TR26.
2. *Metals handbook of machining*, Vol. 3, American Society for Metals (1967).
3. MERCHANT, M. E. (1950). *Machining theory and practice*, American Society for Metals; (1952). *Manual of cutting of metals*, A.S.M.E.
4. KEISSLING, R. (1968). *Non-metallic inclusions in steel*, Part III, Iron and Steel Institute, London, 115.
5. KEANE, D. M. (1974). The influence of inclusions on machinability, *Proc. Conf. on Inclusions and Their Effect on Steel Properties*, I.S.I., Leeds, UK.
6. PALIWODA, E. J. (1964). The role of oxygen in free cutting steels, *Proc. A.I.M.E. Mechanical Working of Steel*, **26**, 27–47.
7. CRAFTS, W. & HILTY, D. C. (1953). Sulphide and oxide formation in steel, *Proc. Electrical Furnace Steel Conf.*
8. SHAW, M. C. *et al.* (1959). The influence of lead on metal cutting forces and temperatures, *Trans. A.S.M.E.*, **79**, 1143–53.
9. STODDART, C. H. *et al.* (1975). Lead monolayer lubrication in steel machining studied by Auger spectroscopy, *Nature*, **253**(5458), 187–9.
10. JOSEPH, R. A. & TIPNIS, V. A. (1975). The influence of non-metallic inclusions on the machinability of free machining steels, In: *The influence of metallurgy on machinability*, A.S.M.
11. WICKER, A. (1967). US Patent 3,301663, Jan 31.
12. KONIG, W. (1965). Der Einfloss nichtmetallscher Einfüsse auf Zerspanbarkeit von unlegieslen Baustählen, *Indust Auzliger*, **87**: Part 1 (26) 463, Part 2 (43) 845 and Part 3 (51) 1033.
13. OPITZ, H. & KONIG, W. (1968). Oxides make steel supermachinable, *Machinery*, **75**, 75–8.
14. SATA, T. *et al.* (1969). Report on the machinability of calcium deoxidised steels, *Bull. Jap. Soc. Proc. Eng.*, **13**(1), 1–8.
15. YEO, R. B. G. (1967). The effect of oxygen in resulphurised steels, *Journal of Metals*, **19**(6), June, 29–32; July, 23–7.
16. TIPNIS, V. A. *et al.* (1973). Calcium deoxidised improved machining steels for automotive gears, *Int. Automotive Eng. Conf.*, Detroit, Michigan, S.A.E. Paper No 730115, Jan. 8–12, 1973.

17. ITO, T. *et al.* (1975). Effect of small amounts of lead and sulphur upon chip disposability of calcium deoxidised steels, In: *The influence of metallurgy on machinability*, A.S.M.
18. TIPNIS, V. A. (1970). *Influence of MnS bearing inclusions on flow and fracture in a machining shear zone*, Soc. Man. Engrs. A.S.M., MR70-713.
19. AKHTAR, S. & MILLS, B. (1971). Unpublished research.
20. BOULGER, F. W. (1958). *Influence of metallurgical properties on metal cutting*, ASTME, MR58-138, 19.
21. GARVIN, H. W. & LARRIMORE, R. M. (1964). Metallurgical factors affecting machining of free machining stainless steel, *Met Soc AIME, Mechanical Working of Steel*, **26**, 133.
22. KOVACH, C. W. & MOSKOWITZ, A. (1970). Modification adds machinability to type 303, *Met. Progress*, **98**(1), 105.
23. KOVACH, C. W. & MOSKOWITZ, A. (1969). Effects of manganese and sulphur on the machinability of martensitic stainless steels, *Trans. AMS AIME*, **245**, 2157.
24. GOLDSHTEIN, Y. E. *et al.* (1973). Effect of selenium on the structure and properties of sulphur steels with high machinability, *Met. Sci. Heat Treat.*, **15**, 286.
25. CLARK, W. C. (1963). Which free machining stainless steels?, *American Mach. Metalworking Mfg.*, **107**, 55.
26. BELOV, A. D. *et al.* (1970). Machinability of cast stainless steel alloys, *Machines and Tooling*, **41**(5), 38.
27. TRENT, E. M. (1977). *Metal cutting*, Butterworths, London.
28. ZLATIN, N. & FIELD, M. (1973). In: *Titanium science and technology*, Vol. 1 (Eds R. I. Jaffe and H. M. Burte), Plenum Press, New York, 409.
29. STODDART, C. T. H. *et al.* (1979). Relationship between lead content of Cu–40% Zn, machinability and swarf surface composition determined by Auger electron spectroscopy, *Metals Technology*, **6**(5), 176–84.

Chapter 6

THE ISO MACHINABILITY TEST

6.1. INTRODUCTION

In Chapter 3 it was mentioned that the specification for a standard long absolute machinability test became available in 1977—this is embodied in standards ISO 3685–1977 and BS 5623:1979. In this chapter a brief resumé is given of the main recommendations of the ISO 3685–1977 standard in an attempt to familiarise the reader with the concepts of the machinability test standards. For those who intend to carry out machinability testing to the standard specification it is recommended that a copy of the full standard be obtained from the appropriate body.

The aim of the standard is to ensure, as far as possible, that future machinability testing of the type described should be carried out using standard conditions for the various variables so that the information obtained may be used to generate compatible comprehensive machinability data arising from many sources. Clearly, much thought has gone into the choice of recommended conditions for testing in an attempt to adequately cover a wide range of practical cutting conditions. The main body of the ISO 3685–1977 report is contained in several sections which deal with reference workpiece materials, reference tool materials and tool geometries, reference cutting fluids, reference cutting conditions, tool life criteria and tool wear measurement, equipment, test procedures, and the recording and reporting of results. Additionally, the standard contains appendices which expand on certain aspects of the standard.

6.2. REFERENCE WORKPIECES

In all tests where the workpiece itself is not the test material a reference medium carbon steel or cast iron should be used depending on which is the most appropriate. The workpiece chemical composition and heat treatment are specified in the standard as is the range of hardness of the workpiece which is acceptable. The workpiece should be free from scale and should be mounted between chuck and centre or between centres. If the length to diameter ratio is such that chatter occurs, testing should be discontinued and the workpiece geometry should be suitably modified. A workpiece length to diameter ratio of greater than 10 to 1 is not recommended. The hardness of the workpiece should be determined over the full cross-sectional area and testing should only be carried out at those diameters where the hardness lies in the specified range.

6.3. REFERENCE TOOL MATERIALS AND TOOL GEOMETRIES

If the tool material itself is not the test variable then it should be one of five basic types: a high-speed-steel tool; an ISO P30 grade cemented carbide; an ISO P10 grade cemented carbide; an ISO K20 grade cemented carbide; or an ISO K10 grade cemented carbide. The high-speed-steel tool should be of specified chemical composition and heat treatment and should have a hardness and grain size in a given range. If cemented carbides are to be used then the P10 and P30 grades should be used for machining steels and the K10 and K20 grades for machining cast iron. Where possible, the reference tool material should be purchased from a tool materials bank. Alternatively, because carbide grades of the same ISO groups will vary between tool manufacturers and, to a lesser extent, between batches, the performance of one of a batch from a particular manufacturer should be calibrated against a standard tool of the same grade.

Because there is no standard for ceramic tools, if these are to be used, the composition and physical properties should be described in as much detail as is possible.

For high-speed-steel tools there is only one recommended geometry, for cemented carbide tools there are two, and for ceramic tools there is one. The geometries are typical of those which would be used in practice.

Tools should be correctly set on the machine, i.e. the corner of the tool should be on centre and the tool shank should be perpendicular to the axis of rotation of the workpiece. Standard tool-holders of specified size should be used and tolerances on tool angles and surface finish should comply with those recommended.

A chip breaker should not be used with high-speed-steel tools unless the chip breaker itself is the test variable or cutting conditions become dangerous. With carbide or ceramic tools a ramp-type chip breaker of specified geometry is permissible and the chip breaker distance should be chosen to give effective chip breaking.

6.4. REFERENCE CUTTING FLUIDS

If the cutting fluid is not the test variable then cutting should be carried out either dry or using the *reference* cutting fluid with a strong recommendation for cutting dry. If a cutting fluid is considered to be necessary, the fluid flow should not be less than a specified amount and the fluid should flood the active part of the cutting tool.

6.5. CUTTING CONDITIONS

Only four sets of values for feed, depth of cut, and tool corner radius are recommended for tests where none of these are the test variable—these are intended to cover conditions from light to heavy roughing operations. When more than one of the above are the test variables it is recommended that only one be changed at a time and that, in any case, the minimum depth of cut should be at least twice the corner radius, the maximum depth of cut should be no more that ten times the feed, and the maximum feed should be no more than 0·8 times the corner radius.

The cutting speed should be measured on the uncut workpiece surface and NOT on the surface resulting from the cut. At least four different cutting speeds should be chosen for each test and ideally the speeds should be such that the resulting tool lives are approximately 5, 10, 20 and 40 min. If testing time must necessarily be reduced or the workpiece material is very expensive, a shorter set of tool lives can be used but the minimum tool life should not be less than 2 min. The tests should be repeated three times for each cutting speed and testing should be done on a pseudo-random basis to minimise the effect of gradual changes in workpiece properties.

6.6. TOOL LIFE CRITERIA AND TOOL WEAR MEASUREMENTS

The type of wear which first reaches the value of tool wear considered to be the limit for that type of wear should be used to determine the useful life of the cutting tool. The type of wear and the tool life criterion should be reported. If it is not clear which type of wear will predominate, all relevant wear measurements should be taken. In some circumstances the criterion will change with changes in cutting speed and this will result in a broken cutting speed–tool life curve as shown in Fig. 6.1. For high-

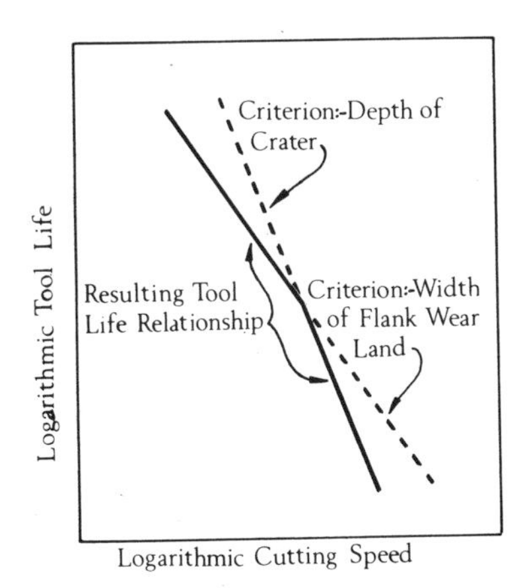

FIG. 6.1. Broken cutting speed–tool life curve for combined flank and crater wear.

speed-steel tools three criteria of tool failure are usually used and these are:

(1) Catastrophic failure;
(2) If the flank wear is even, an average flank wear land width of 0·3 mm; and
(3) If the flank wear land is irregular, scratched, chipped or badly grooved, a maximum flank wear land width of 0·6 mm.

Of these, by far the most common criterion is that of catastrophic failure.

For cemented carbide cutting tools three criteria of tool failure are usually used and these are:

(1) If the flank wear is even, an average flank wear land width of 0·3 mm;
(2) If the flank wear land is irregular, scratched, chipped or badly grooved, a maximum flank wear land width of 0·6 mm; and
(3) A crater depth of $(0{\cdot}06+0{\cdot}3f)$ mm where f is the feed in millimetres per revolution.

Of these, by far the most common criterion is flank wear and usually an average wear land of 0·3 mm. The general exception to this is machining cast irons at high speed when, often, the tool failure mode is cratering.

For ceramic tools three criteria of tool failure are normally used and these are:

(1) Catastrophic failure;
(2) If the flank wear is even, an average flank wear land width of 0·3 mm; and
(3) If the flank wear land is irregular, scratched, chipped or badly grooved, a maximum flank wear land width of 0·6 mm.

The various types of wear are illustrated in Fig. 6.2.

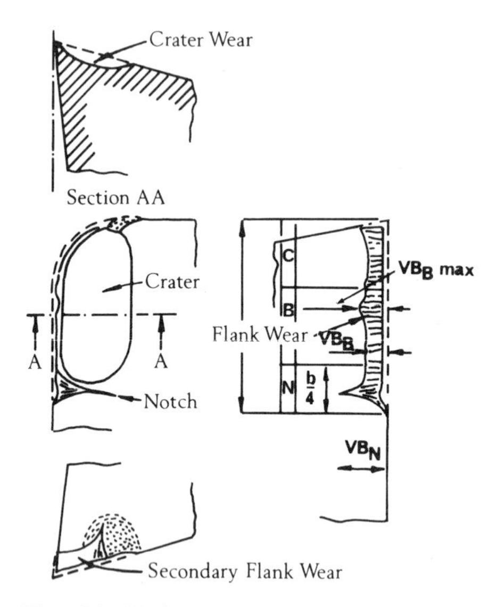

FIG. 6.2. Major types of cutting tool wear.

Other wear phenomena, such as notch wear, wear of the minor flank, plastic deformation of the tool corner, and edge chipping may occur in practise but all of these eventually result in one of the preferred criteria being valid and this criterion should be used. In the unusual case of premature failure of the tool which is invariably caused by a 'hard spot' in the workpiece material, a machine malfunction or unduly severe cutting conditions, tool failure is unpredictable and values obtained should NEVER be used as a measure of tool life.

6.7. TOOL WEAR MEASUREMENT

Parts adhering to the flank directly under the wear land can give the appearance of a large width to the wear land. Also, a deposit in the crater results in lower values of the crater depth. Loose material should be removed carefully but chemical etchants should not be used except at the end of the test. For the purpose of the wear measurements the major cutting edge is considered to be divided into three zones as shown in Fig. 6.2:

Zone C is the curved part of the cutting edge at the tool corner;
Zone N is the quarter of the worn cutting edge length *b* farthest away from the tool corner; and
Zone B is the remaining straight part of the cutting edge between zone C and zone N.

The width of the flank wear land VB_B should be measured within zone B in the tool cutting edge plane perpendicular to the major cutting edge. The width of the flank wear land should be measured from the position of the original major cutting edge. The crater depth KT should be measured as the maximum distance between the crater bottom and the original face in zone B.

6.8. EQUIPMENT

The test lathe should be in good condition and have an infinitely variable speed drive which covers the range of cutting speeds to be used. Other equipment should include:

(1) Equipment for measuring tool geometry accurately to ensure that it conforms to the tolerances specified for the test;
(2) A stop-watch to measure cutting time;
(3) A suitably calibrated microscope, preferably a travelling microscope;
(4) A suitable dial indicator for measuring crater depth or, preferably, measuring equipment to measure the crater profile;
(5) Hardness measuring equipment;
(6) A tachometer or other suitable method of measuring cutting speed;
(7) Equipment for measuring workpiece diameter; and
(8) Equipment for measuring the flow rate of the cutting fluid and the surface finish of the workpiece where necessary.

6.9. TOOL LIFE TEST PROCEDURE

Tool wear measurements should be made at suitable intervals. The interval should be short while primary wear is taking place and then generally be increased to appropriate lengths for secondary wear measurement depending on the rate of wear. In any case each tool wear–time curve should contain at least five points with, preferably, five points for results in the secondary wear region. Under no circumstances should tool life be determined by extrapolating the tool wear–time curve.

The data should be recorded on a standard sheet which should also include the graph of tool wear versus time. The results of a series of tests should be recorded on a standard data sheet which also includes the cutting speed versus tool life graph.

6.10. EVALUATION OF TOOL LIFE DATA

In the unusual circumstance when only a single cutting speed can be used, the tool life can be expressed in minutes or alternatively as the number of workpieces produced. The cutting conditions, tool geometry, workpiece material and geometry, etc. should be fully specified. In the more general case the constants of the Taylor tool life equation, $VT^{-1/k}=C$, can be estimated from the cutting speed versus tool life graph either by fitting the best straight line to the graph 'by eye' or by using a regression analysis. If a regression analysis, a measure of dispersion, is used significance and confidence interval limits can be determined for the cutting speed–tool life data.

Chapter 7

THE EFFECT OF MACHINABILITY DATA ON METAL REMOVAL PERFORMANCE AND ECONOMICS

7.1. INTRODUCTION

Chapter 3 describes some of the many types of machinability test which have been, and continue to be, used commercially and undoubtedly the widest use of machinability testing, as opposed to the use of data obtained from machinability testing, is to compare the performance of a tool–workpiece combination with that of a reference tool–workpiece combination using the same tool in order to predict whether the expected performance is likely to be better than, equal to, or worse than the reference standard. Based on this logical yes/no situation, a decision can be made regarding the probability of the use of the tool–workpiece combination being appropriate, i.e. with a knowledge of past experience of the test and decisions based on the results of tests, a level of confidence in the test can be established which in turn results in an indication of the probability of the tool–workpiece combination being capable of meeting the requirement satisfactorily. In most metal cutting processes, success is measured by the ability of the cutting tool to cut effectively for a given period of time for given cutting conditions or, more commonly, to carry out the required operations on a given number of workpieces.

As was discussed in Chapter 3, many of the common ranking tests are quick and effective for this type of work but the amount of information which is obtained and stored is minimal. Thus, while this type of test is usually excellent for judging against a standard it is in no way of any use to judge the effectiveness of the standard. The effectiveness of the standard can only be judged by reference to a variety of parameters, which includes the relationship between cutting conditions and tool life, which can only be determined from an absolute machinability test.

Before discussing how this machinability data is used, it is worthwhile considering the factors which influence the effectiveness with which a cutting operation is carried out.

Usually, an acceptable workpiece is produced when it is geometrically correct within the specified tolerances and the surface finish produced meets a given specification. Sometimes, the nature of the surface is also important but since this is only relevant in a relatively small proportion of all metal cutting operations this aspect of metal removal will not be considered. It is, of course, important to ensure that both the tolerance and the surface finish requirements are as generous as possible since tightening tolerances and improving surface finish for a given process invariably result in an increased cost; this design aspect of metal removal processes is outside the scope of this text.

The conventional method of producing a workpiece which is geometrically acceptable is to:

(1) Remove the bulk of the excess material at high metal removal rates (roughing) where, due to the lack of machine tool and tool stiffness, the workpiece geometry will not be that which was nominally intended and where the surface finish will generally be worse than that required; and
(2) Remove the remaining small amount of material under conditions where the forces acting on the tool will be low (finishing) and where the required surface finish can be obtained.

Most absolute machinability data is generated using bulk metal removal cutting conditions and this is important in that severe tool weakening due to wear is most likely to result in a major tool failure when the forces acting on the tool are high. However, the main effect of tool wear in practice is to produce workpieces which are oversize and which have a poorer surface finish than would be obtained with an unworn tool. Both these effects are more pertinent to finishing rather than roughing operations and it is important, therefore, that undue emphasis should not be placed on bulk metal removal.

In theory, comprehensive absolute machinability data would contain at least the values of the exponents p, q and r of Taylor's extended tool life equation

$$V^{\mathrm{p}} S^{\mathrm{q}} a^{\mathrm{r}} T = \mathrm{C}_2 \qquad [7.1]$$

where V is the cutting speed, S is the feed, a is the width of cut, T is the tool life, and C_2 is a constant. In practice, machinability data banks often present limited information which is 'typical' of a reasonable range of practical cutting conditions. Thus, even though it is sometimes unwise, machinability data can be interpolated to give results for specific conditions. In particular, if techniques such as the use of chip equivalent concepts are employed, a good estimate of the effect of the main cutting parameters on tool life can be obtained. Although it is not common to be given the values of p, q and r of Taylor's equation a knowledge of the range of values of these and how they compare with each other for common tool–workpiece combinations is important when determining a strategy for metal removal.

Table 7.1 shows a typical range of values for p, q and r for high-speed-steel and cemented carbide tools when used to cut a variety of steels in

TABLE 7.1

TYPICAL RANGE OF VALUES FOR p, q AND r FOR TAYLOR'S EXTENDED TOOL LIFE EQUATION $V^p S^q a^r T = C_2$ [a]

Tool	p	q	r
High-speed-steel	4 to 7	1 to 2	0 to 1
Cemented carbide	2 to 4	0·5 to 1·5	−0·5 to 0·5

[a] V=cutting speed (m min^{-1}); S=feed (mm rev^{-1}); and a=width of cut (mm).

turning. Two effects may be observed from the table:

(1) The exponent p related to cutting speed is very much larger than the exponent q related to feed which, in turn, is generally larger than the exponent r related to width of cut, i.e. tool life will be least affected by changes in width of cut and most affected by changes in cutting speed. Thus, since metal removal rate is directly related to cutting speed, feed, and width of cut, it follows that within restrictions imposed by the process, the method for obtaining high metal removal rates with small amounts of tool wear is to increase the width of cut to as large a value as is possible. This should be followed by increasing the feed to as large a value as is possible and finally the cutting speed should be adjusted to a value which gives optimal cutting for the chosen cutting criterion. In practice, in turning, for example, the depth of cut is usually limited by the amount which can be removed without making the workpiece under size or by difficulties of controlling the chip. The feed is restricted by the force which the tool, the tool-holder or, more commonly, the machine tool can withstand without failure of the

tool due to direct forces or vibration. For given values of width of cut and feed, the cutting speed is ultimately restricted by the power which is available at the spindle of the machine although, as will be seen later, it is more common for the maximum speed of the machine to be too low.

(2) The average of the range of values of p for cemented carbide is much smaller than the average of the range for high-speed steel and the value of C_2 is much larger. Thus, for all but very low cutting speeds, not only is the tool life greater for a cemented carbide than for a high-speed-steel cutting tool but, as the cutting speed is increased, the tool life gets proportionately larger for the cemented carbide even though, of course, the tool life is reducing.

For finishing operations, metal removal rates are necessarily low since, by definition, the width of cut will be small and the feed will be restricted by the requirements of surface finish. Although geometric surface finish (that obtained without taking into account inefficiencies in the cutting process) is a function of feed and tool geometry only, the actual surface finish, which is usually worse than geometric surface finish, is also a function of cutting speed. With the exception of very low cutting speeds, where unstable built-up-edges are formed, an increase in cutting speed results in the actual surface finish tending to the geometric surface finish. Consequently, if the tool material is cemented carbide where the optimum cutting speed for any reasonable criterion of machine performance is high, calculating the value of feed to give the required geometric surface finish and reducing this by a small percentage will generally result in an acceptable surface finish. However, if the tool material is high-speed steel then the optimum cutting speed will be relatively low and this will have a significant effect on the surface finish and the feed may have to be reduced considerably from that which would give an acceptable geometric surface finish.

7.2. CRITERIA OF PERFORMANCE

Several criteria can be used as a basis for the successful operation of a machine tool. The three most common are: (1) a minimum cost criterion; (2) a maximum production rate criterion; and (3) a maximum profit criterion.

Whichever criterion is chosen, two factors are important when scheduling work to a machine: the cutting conditions under which the machine operates; and the range of the number of machined components that the machine can economically produce (the economic batch size).

The effects of cutting condition have already been described and, later in this chapter, simple analyses will show how the optimum cutting speed for minimum cost and maximum production can be determined for single point turning, and that these are independent of the number produced. The economic batch size is controlled by the cutting conditions, the machinability, and the type of machine to be used, where the significant parameter is the ratio of how much it costs to set up a machine compared with how fast the machine can produce parts—this will also be dealt with later.

Whatever criterion of performance is chosen, it is significant that the optimum cutting conditions do not depend on the size of the batch being produced and this is important since it means that the performance of an individual machine is unaffected by other production conditions and whilst, because of these other factors, a particular machine may not be capable of producing economic batches from those put to the machine, it should always be capable of removing metal in an optimal manner.

When choosing the criterion of performance it is easy to confuse the effects of each criterion. In theory, there could be a considerable difference between cutting conditions for minimum cost and maximum production but in practice the cutting conditions for these two criteria are often very similar. A maximum profit criterion will depend on both the production rate and the cost of production and this will be discussed later.

7.3. ECONOMICS OF TURNING OPERATIONS

For simplicity the analyses shown will be for single point turning and an examination will be made of three separate types of turning machine—the centre lathe, the capstan and the numerically controlled (NC) lathe. For each type of machine, turning will be considered using a high-speed-steel cutting tool and a cemented carbide cutting tool, and an appropriate value of feed will be chosen for each type of tool and batch sizes of 1 to 10 000 will be considered. In each case, the cutting speeds for minimum cost production and maximum production rate will be determined and the overall cost of producing the various batch sizes for each type of machine, for each type of tool and for each criterion of economic production will be calculated. Again, for simplicity, it will be assumed that the cutting operation will be axial turning of cylindrical stock and that the only part of the process to be considered will be the axial turning element, i.e. no acount will be taken of the parting off operation. The figures used are typical of the times and costs for the operators, the equipment and the labour, but even if these are subject to some errors they will be accurate enough to demonstrate the various effects attributable to the type of equipment, the type of tool and the batch size.

For single point turning, times can be attributed to batches, parts and incidence of tool changes. The set-up time, T_s, occurs once per batch. Three times can be related to the parts: the time to load and unload the part, T_L; the time to advance and withdraw the tool, T_A; and the time to machine the part, T_M. In addition to these times will be the time to change a tool (in the case of a high-speed-steel tool) or index an insert (in the case of a cemented carbide tool), T_I.

In order to obtain the costs of the activities concerned with the above times, extra data is required. It will be assumed that the running costs of the machine—which include the cost of the operator, the cost of the machine and the cost of the overheads—are M per minute and that the average cost per regrind (high-speed-steel tool) or cost per edge (cemented carbide insert) is C_T.

If it is now assumed that a batch of N components will be produced and that P parts per tool change (insert index) can be manufactured, the cost of the various elements which make up the total cost of producing a batch of N parts can be determined.

The cost of setting up, loading and unloading will be $M(N(T_L + T_A) + T_s)$, the cost of tool replacement will be $M(T_I + C_T)N/P$, and the cost of machining will be MNT_M.

If Taylor's simple tool life equation is taken to be valid, then $VT^{-1/k} = C_1$, where V is the cutting speed, T is the tool life, and k and C_1 are constants. Further, since the cut length for axial turning is $\pi DL/S$, where D is the diameter of the bar, L is the length, and S is the feed per revolution, the time to machine a part, T_M, will be $\pi DL/SV$, and the number of parts produced per tool change (insert index), P, will be T/T_M, i.e.

$$P = \frac{C_1^{-k} SV}{V^{-k} \pi DL} \qquad [7.2]$$

Thus, the labour cost of tool changing and the capital cost of the tools to produce a batch of N parts will be $NV^{-k-1}\pi DL(MT_I+C_T)/C_1^{-k}S$. The machining cost of producing a batch of N parts will be MNT_M i.e. $MN\pi DL/SV$, and, when this is added to the set-up costs and the tool and tool change costs, the total cost, C_{TOT}, of producing a batch of N parts will be given by

$$C_{TOT}=M(N(T_L+T_A)+T_s)+\frac{NV^{-k-1}\pi DL}{C_1^{-k}S}(MT_I+C_T)+\frac{MN\pi DL}{SV} \qquad [7.3]$$

If this is now differentiated with respect to V, and equated to zero, the cutting speed for minimum cost, V_c, can be determined, and is given by

$$V_c=C_1\left(\frac{M}{(MT_I+C_T)(-k-1)}\right)^{-1/k} \qquad [7.4]$$

As mentioned previously, the optimum cutting speed for minimum cost is independent of the batch size but it is of interest to examine the factors which do influence it.

Clearly, if either or both the time to change cutting tools and the cutting tool cost are reduced, the optimum cutting speed will increase. If the constant C_1 and/or the machine and labour costs are increased the optimum cutting speed will increase and, perhaps surprisingly for practical conditions, altering k has very little effect. Thus, if expensive machine tools are to be used and labour becomes more expensive with other relevant factors remaining unchanged, the economic cutting speed will rise. It becomes important therefore when operating numerically controlled machines, for example, to ensure that the machine tool is capable of achieving the cutting conditions necessary to operate it in an economic manner. In inflationary times if machine and labour costs rise it could be expected that cutting tool costs will also rise and this factor tends to reduce the economic cutting speed. However, evidence of what has happened in the last 10 years would suggest that cutting tools have become effectively cheaper and consequently economic cutting speeds should have increased.

It is even more revealing to examine the effect on economic cutting speed of improvements in tool performance over the years. For high-speed-steel tools typical values of k and C_1 would be 5 and 80 respectively whereas for cemented carbide tools these would be 2·5 and 300. Further, the cost of a cutting edge for a cemented carbide tool is very much less than the equivalent edge for a high-speed-steel tool. Table 7.2 shows typical values for M, T_I, C_T, T_s, T_L, T_A, k and C_1 for high-

TABLE 7.2
TYPICAL COST AND TIME PARAMETERS FOR TURNING

		Cemented carbide	High-speed-steel
k		−2·5	−5·0
C_1		300	80
Centre lathe			
	M	3·5	3·5
	C_T	2·5	10
	T_s	30	30
	T_L	2	2
	T_A	1	1
	T_I	1	2
Economic cutting speed		205·6	44·2
Capstan lathe			
	M	4·5	4·5
	C_T	2·5	10
	T_s	120	120
	T_L	0·1	0·1
	T_A	0·5	0·5
	T_I	1	2
Economic cutting speed		213·8	45·5
NC Lathe			
	M	9	9
	C_T	2·5	10
	T_s	30	30
	T_L	0·2	0·1
	T_A	0·2	0·2
	T_I	1	2
Economic cutting speed		231·3	48·3

Costs are based on a labour cost of unity.
Times are expressed in minutes

speed-steel tools and cemented carbide tools respectively and for a feed of 0·5 mm rev^{-1} together with the economic cutting speed for each. Using the figures from the table it can be seen that, typically, changing from high-speed-steel tools to cemented carbide tools increases the economic cutting speed by a factor in excess of five. If this is compared with the increase in maximum lathe rotational speed since the introduction of cemented carbides, the increase in speed (and power) obtainable has not matched the increase required, i.e. it can be concluded that except for a small proportion of work it is unlikely that a modern machine tool is capable of operating under economic cutting conditions. These arguments are an over-simplification of practice in that no consideration has been given to the facts that high-speed-steel tools are capable of withstanding higher feeds than cemented carbides, there may be swarf control problems associated with operating at high cutting speeds and on non-automatic machines safety features may not be adequate to deal with high-speed cutting. However, even after allowance has been made for these factors the general conclusion outlined above is valid. Cook (1) quotes an example where he compares the cost as obtained from a minimum cost analysis with that obtained using 'standard' data as supplied by the *Tool engineers handbook*. For this example, the optimised cost is one-tenth of that obtained using the standard data and Cook concluded that, based on this type of comparison and on the experience of seeing machine tools under production conditions, the majority of cutting operations are being carried out under conditions which are at least 50% more costly than is necessary. This observation serves to emphasise the importance of reliable and up-to-date machinability information. However diligent the subscribers to information centres, such as the *Tool engineers handbook*, are it is certain that much of the machinability data provided by this type of source will be suspect because of improvements in the performance of both cutting tools and workpiece materials. It could be expected that where the information required is available, if it is obtained from a data bank, it will be relevant. However, for a situation where substantial amounts of machining are required using workpiece and/or tool materials for which data is either not available or suspect, it is recommended that the user commission a machinability study or carry out appropriate tests to determine the information.

If the cutting speed for maximum production is required then the total production time, T_{TOT}, is given by

$$T_{TOT}=N(T_L+T_A)+\frac{T_s+NV^{-k-1}\pi DLT_I}{C_1^{-k}S}+\frac{N\pi DL}{SV} \qquad [7.5]$$

and if this is differentiated with respect to V and equated to zero the cutting speed for maximum production, V_p, will be given by

$$V_p=C_1\left(\frac{1}{T_I(-k-1)}\right)^{-1/k} \qquad [7.6]$$

If the cutting speed for maximum production is now divided by the cutting speed for minimum cost this gives

$$\frac{V_p}{V_c}=\left(1+\frac{C_T}{MT_I}\right)^{-1/k} \qquad [7.7]$$

which is always greater than unity, i.e. the cutting speed for maximum production is *always* greater than the cutting speed for minimum cost. From the figures used previously for C_T, M, T_I and k for high-speed-steel tools and cemented carbide tools and, say, the centre lathe, V_p/V_c for high-speed-steel tools is 1·194 and for cemented carbide tools is 1·241. Thus, the difference in the ratio of the cutting speeds for maximum cost and minimum production for high-speed-steel tools and cemented carbide tools is relatively small but, of course, the actual cutting speeds are very different.

Although the cutting speeds for both minimum cost and maximum production are unaffected by the batch size, the batch size will influence the choice of machine on which the part should be manufactured. Traditionally and before the introduction of numerically controlled machines it was accepted that, in turning, small batches would be manufactured on centre lathes, medium-sized batches would be best suited to capstan and turret-type lathes, and large batches could be most economically produced on automatic lathes. As a generality this was true, but even though the cutting speeds for minimum cost and maximum production are unaffected by batch size, the machinability of the tool–workpiece combination does significantly affect the economic batch sizes for the various types of turning machine and this will be illustrated.

The effect of cutting speed on cost and production rate for the data given in Table 7.2 and for batch sizes of 1100 and 10000 is shown in Figs 7.1 to 7.6. From these figures, apart from the already mentioned

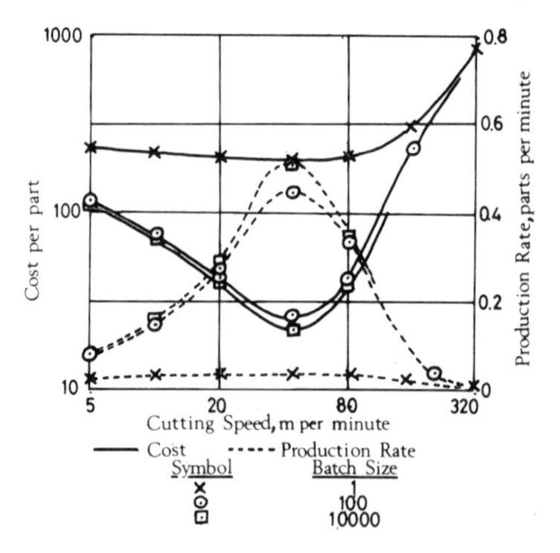

FIG. 7.1. Tool: high-speed-steel; Machine: centre lathe.

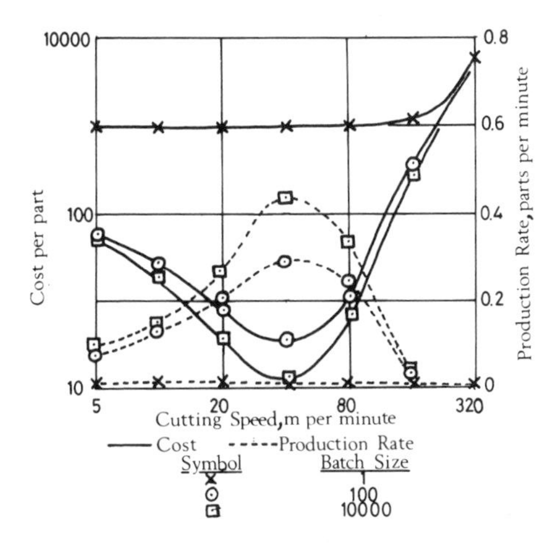

FIG. 7.2. Tool: high-speed-steel; Machine: capstan lathe.

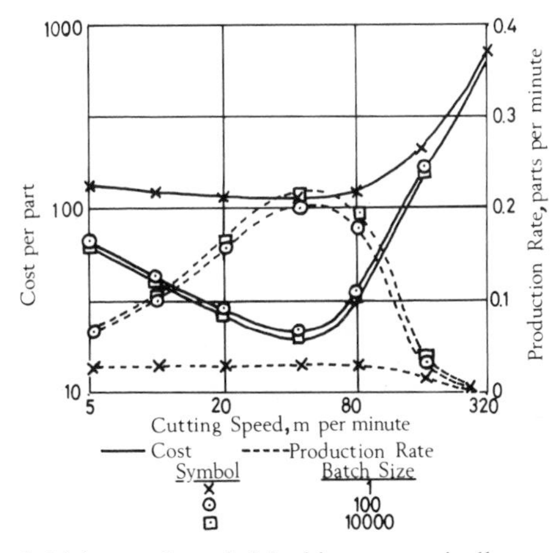

FIG. 7.3. Tool: high-speed-steel; Machine: numerically controlled lathe.

independence of optimum cutting speed from batch size for both minimum cost and maximum production rate, two more effects can be observed—these are:

(1) The rate of change of both cost and production rate with changes in cutting speed is much greater for high-speed-steel tools than for cemented carbide tools, i.e. it is far more important to operate near to the optimum speed for high-speed-steel tools than for cemented carbide tools if costs are to be minimised and production rates

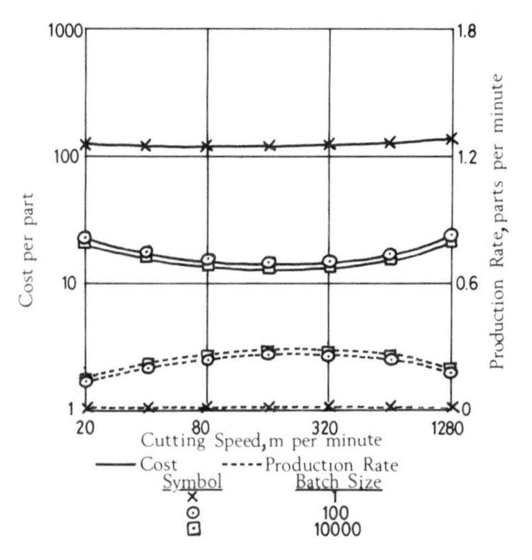

FIG. 7.4. Tool: cemented carbide; Machine; centre lathe.

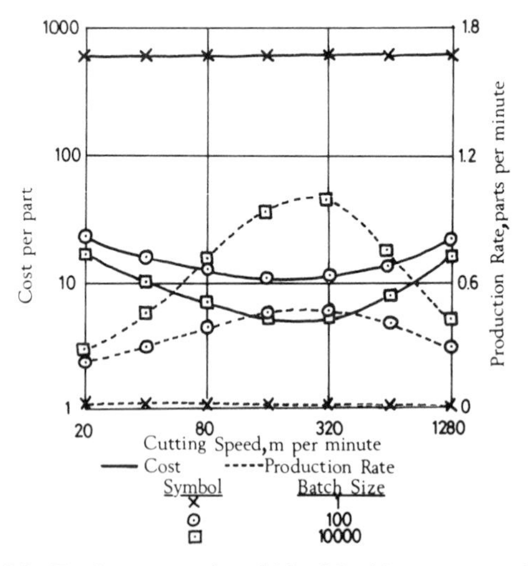

FIG. 7.5. Tool: cemented carbide; Machine: capstan lathe.

maximised. Fortunately, since optimum cutting speeds for high-speed-steel tools are generally achievable on lathes, this effect should not prove to be inhibiting although this, of course, presupposes that the necessary information about the tool, the work-piece and the machine is available and that the appropriate analyses have been performed.

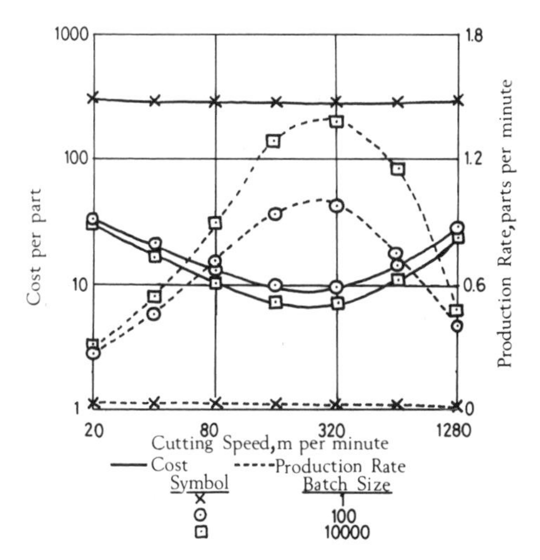

FIG. 7.6. Tool: cemented carbide; Machine: numerically controlled lathe.

(2) Whichever type of tool is used, increasing batch size has a reducing effect on the performance of a centre lathe at a much smaller batch size than for either the capstan or numerically controlled lathes. This is a result of two factors: firstly, in the case of the capstan lathe, the large set-up times and the associated costs still have a significant effect on cost and production rate for relatively large batch sizes; and secondly, for both the capstan and the numerically controlled lathes, the non-productive time associated with the manufacture of each part is much smaller than that for a centre lathe.

7.4. MACHINING FOR MINIMUM COST

Consideration is now given to the effect on cost of production of changes in batch size, type of machine, type of tool, and feed when the machine is performing at the economic cutting speed. For the purpose of these analyses it has been assumed that an extended version of Taylor's tool life equation of the form $V^p S^q T = C_3$ is applicable where $q = 1$ for both high-speed-steel and cemented carbide tools, and $p = 5$ and 2·5 for high-speed-steel and cemented carbide tools, respectively. If it is also assumed that, for a feed of 0·5 mm rev^{-1}, the cutting speed for a tool life of 1 min for high-speed-steel is 80 mm min^{-1} and for cemented carbide is 300 m min^{-1} then the values of C_1 in the simpler Taylor equation ($VT^{-1/k}=C_1$) will be as given in Table 7.3.

TABLE 7.3
MINIMUM COST MACHINING DATA

S	C_1	*Centre lathe* V_c	*Capstan lathe* V_c	*Numerically controlled lathe* V_c
High-speed-steel tool				
0·125	105·6	58·34	60·00	63·78
0·5	80·0	44·20	45·45	48·32
2·0	60·6	33·48	34·43	36·60
Cemented carbide tool				
0·125	522·3	357·97	372·16	402·63
0·5	300·0	205·61	213·76	231·26
2·0	172·3	118·09	122·77	132·82

Further, using these data and the data given in Table 7.2 and applying them to the equation (eqn. 7.4) for determining the cutting speed for minimum cost production for each value of feed gives the data also presented in Table 7.3. From this information, the relationship between cost of production and batch size for each feed, each type of machine and each type of tool can be determined. These relationships are shown graphically in Figs 7.7 to 7.9 where it can be seen that:

(1) Increasing the batch size always reduces the cost of production and this is more pronounced for cutting with carbides and for increasing feeds; and

(2) For centre lathe operations, even though the economic cutting speeds for cemented carbides are very much greater than those for high-speed steels, the cost of production is only significantly different for small values of feed. This trend is also observable for both the capstan lathe and the numerically controlled lathe although the effect occurs at much larger values of feed.

This is not a general effect but a function of the ratio of the non-cutting time to the cutting time associated with the production of one part. In this case, where for the centre lathe this value is high, the effect is much more noticeable than, for example, it would be if the job was very much larger and the cutting time was significantly

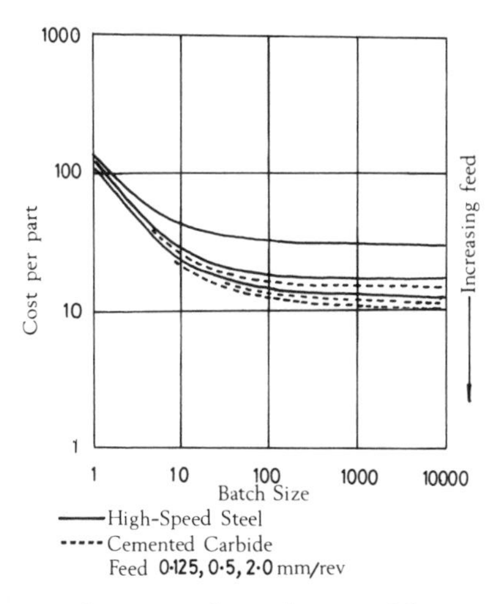

FIG. 7.7. Cost at the economic cutting speed for a centre lathe.

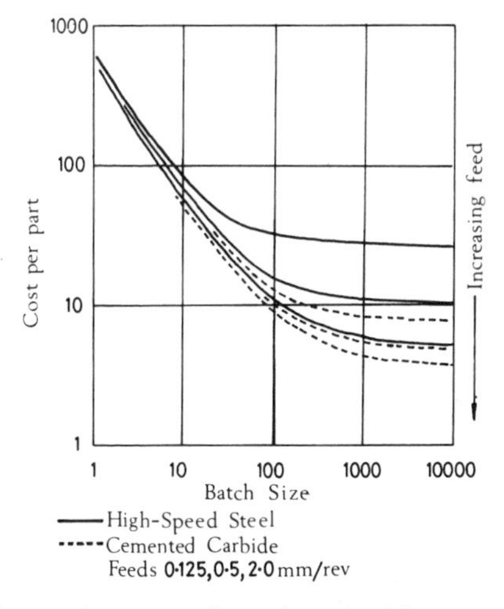

FIG. 7.8. Cost at the economic cutting speed for a capstan lathe.

increased. However, one general observation which can be made is that non-productive time, if excessive, can nullify the potential benefits of cutting at high speeds with cemented carbides and every effort should be made to provide features on the machine which will help to eliminate this.

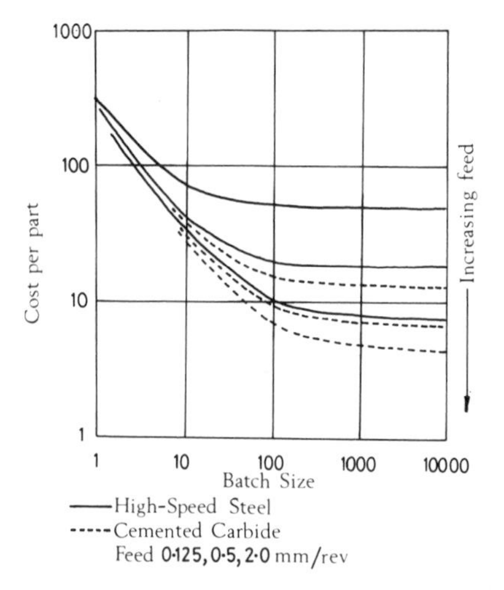

FIG. 7.9. Cost at the economic cutting speed for a numerically controlled lathe.

If the results presented in Figs 7.7 to 7.9 are combined they will give the minimum cost versus batch size relationship for the most appropriate machine for the three values of feed considered—these are shown in Figs 7.10 to 7.12. The important effects which can be observed from these

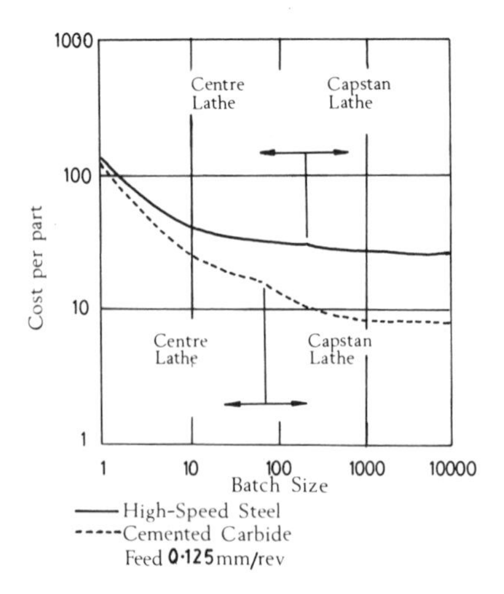

FIG. 7.10. Minimum cost at the economic cutting speed.

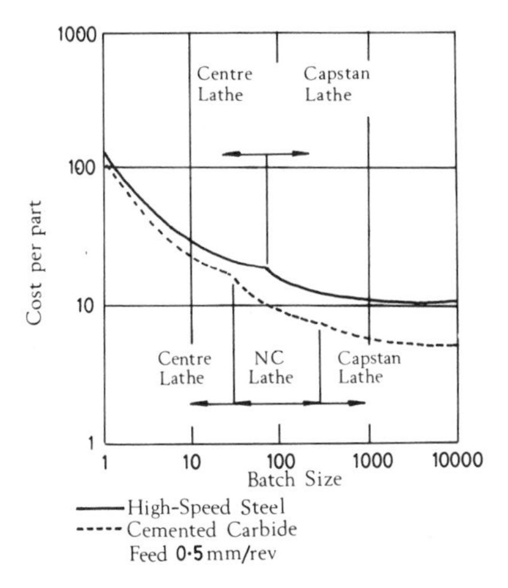

FIG. 7.11. Minimum cost at the economic cutting speed.

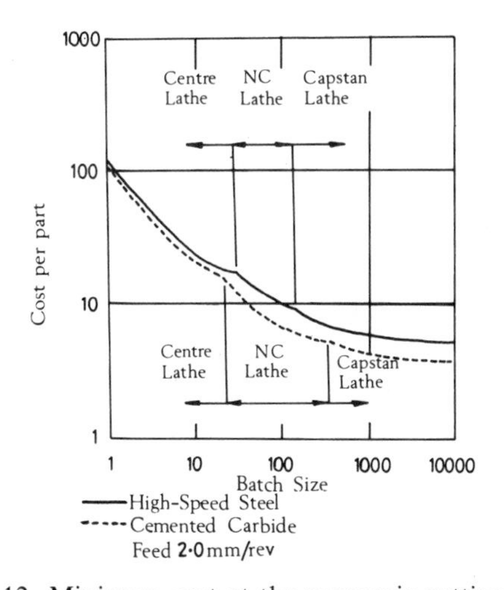

FIG. 7.12. Minimum cost at the economic cutting speed.

relationships are:

(1) The economic batch size for the appropriate machine is dependent on both the feed and the machinability of the work–tool combination. Thus, for a given feed, the economic cutting speed is independent of batch size for a particular machine but the machine which is most appropriate for a given batch size does depend on

the feed. Similarly, for a given work–tool combination, the economic cutting speed is independent of batch size but, again, the machine which is best suited to a given batch size does depend on the machining characteristics of the work–tool combination.

(2) As the feed is reduced or, more significantly, as the cutting time to non-cutting time per part is increased, the benefits of quick tool advance-and-retract time and positioning of numerically controlled machines are not as apparent and, indeed, a situation arises where a numerically controlled machine is not economic for any batch size.

Conversely, as the cutting time is reduced the range of batch sizes for which NC machining is suitable increases whatever the machinability of the work–tool combination but particularly for free-machining applications.

7.5. MACHINING FOR MAXIMUM PRODUCTION

Using the values of feed and the constant C_1 of Table 7.3, the cutting speeds for maximum production for the two types of tool considered are given in Table 7.4. If these data are now inserted into the production rate

TABLE 7.4
MAXIMUM PRODUCTION RATE MACHINING DATA

	S	C_1	V_p
High-speed-steel tool			
	0·125	105·6	69·7
	0·5	80·0	52·8
	2·0	60·6	40·0
Cemented carbide tool			
	0·125	522·3	444
	0·5	300	255·0
	2·0	172·3	146·5

equation (Eqn. 7.5), the relationship between production rate and batch size for the different machines, the different tools and the different feeds can be determined (Figs 7.13 to 7.15). As could perhaps be expected, the results for maximum production rate follow very much the same pattern

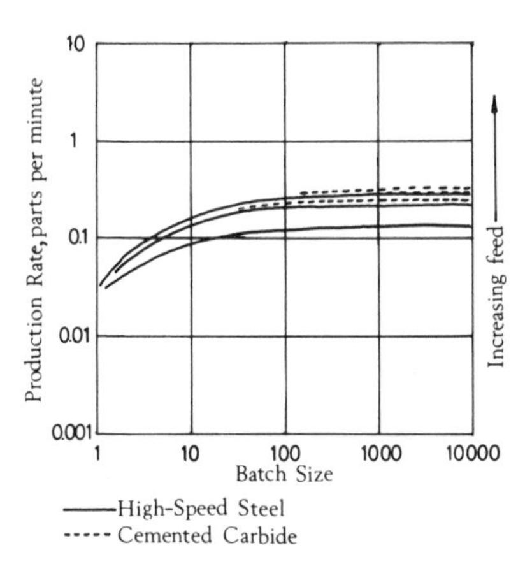

FIG. 7.13. Production rate at maximum production speed for a centre lathe.

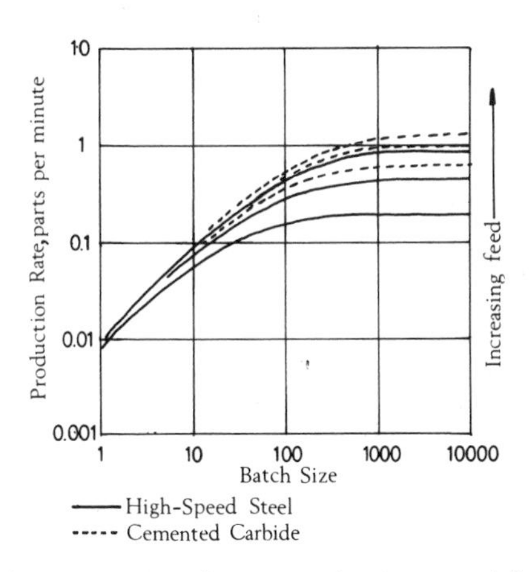

FIG. 7.14. Production rate at maximum production speed for a capstan lathe.

as those for minimum cost and it may be concluded that:

(1) The maximum production rate increases with increasing batch size for all types of machine, feed and tool but the rate of change reduces with increasing batch size;
(2) For a decreasing ratio of cutting to non-cutting time per part, the effect of improving machinability on production rate reduces; and
(3) The maximum production rate depends on the non-cutting time per part and the set-up time and consequently, for this particular

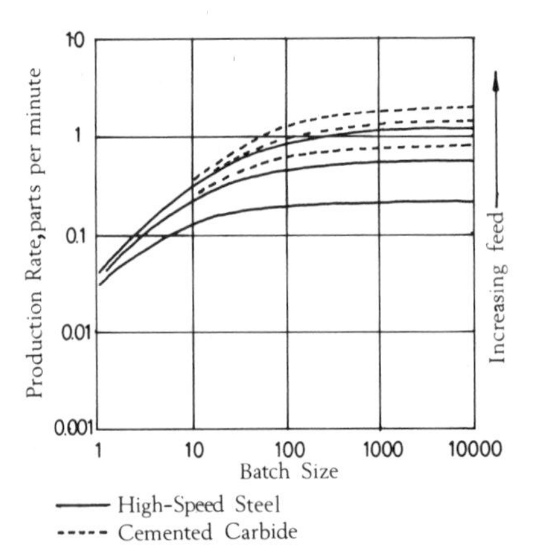

FIG. 7.15. Production rate at maximum production speed for a numerically controlled lathe.

example, the numerically controlled lathe for otherwise identical conditions will always produce at a higher rate than the other two types of machine.

As a result of conclusion (3) above, it is not relevant to determine the machine most appropriate for given batch sizes since in this example the numerically controlled machine is always the best. However, if different data had been used, this would not necessarily have been the case and although data which were considered to be appropriate were chosen there will obviously be circumstances where the significant ratios are such that results similar to those for minimum cost could have been obtained.

7.6. MACHINING FOR MAXIMUM PROFIT

The difficulty of applying a maximum profit criterion is that the 'correct' answer depends on the datum used for measuring profit. To illustrate this two situations have been analysed, one for a high-speed-steel tool and the other for a cemented carbide tool. For each of these the datum from which profit is to be measured is considered to be the cost per minute of cutting at the economic cutting speed on a centre lathe for a batch size of 10 and a feed of 0·5 mm rev^{-1}. With this datum the profit per minute when using each type of machine is based on percentage profits for the centre lathe of 1, 3·162, 10, 31·62 and 100%. The results for the high-speed-steel tool are shown in Figs 7.16 to 7.20 where profit per minute is plotted against batch size for each type of machine. From these the relationship between batch size and percentage profit for the most profitable type of machine can be determined and this is shown in Fig. 7.21 where it can be seen that for small 'profits' the capstan lathe

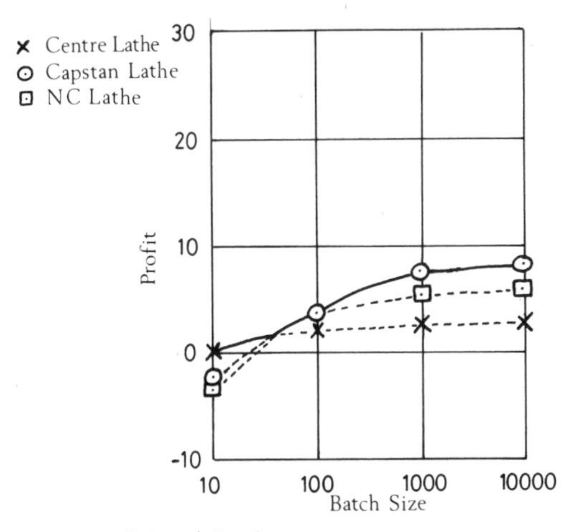

FIG. 7.16. Profit for various types of machine using a profit 'standard' of 1% for a centre lathe producing a batch of 10.

becomes profitable at a relatively small batch size and the numerically controlled lathe is never profitable, whereas for large profits the centre lathe is profitable for only very small batch sizes and the numerically controlled lathe is profitable for all but very large batch sizes. Fig. 7.22 shows the equivalent results for a carbide cutting tool and here, for the criterion chosen, the capstan is never economic and for all but very small batch sizes the numerically controlled lathe is the most profitable machine for all levels of profit.

When Figs 7.21 and 7.22 are compared with Fig. 7.11—the figure showing economic batch sizes for the three types of machine based on minimum cost—the dissimilarities are apparent and lead to the question as to which criterion of performance should be used since this will very often affect the choice of machine. In many ways the maximum profit criterion is very attractive on the basis that the ultimate objective of a machining operation is to do it in a profitable manner. However, the

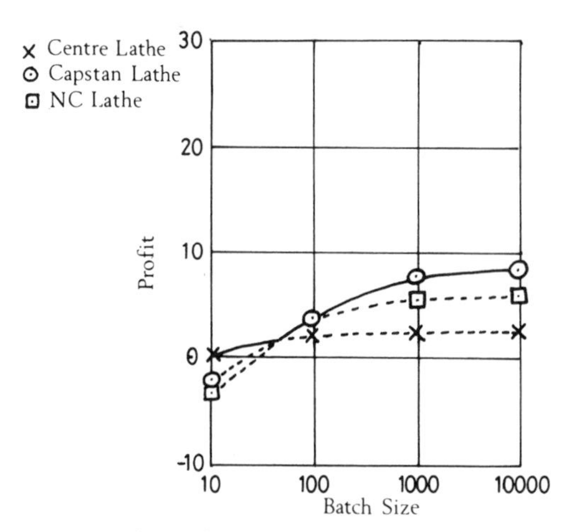

FIG. 7.17. Profit for various types of machine using a profit 'standard' of 3·16% for a centre lathe producing a batch of 10.

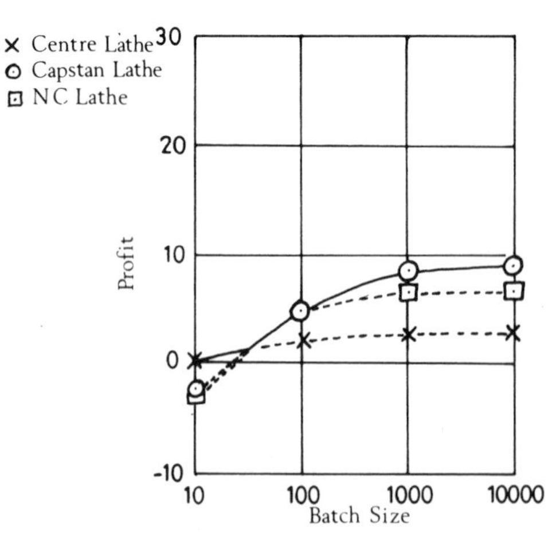

FIG. 7.18. Profit of various types of machine using a profit 'standard' of 10% for a centre lathe producing a batch of 10.

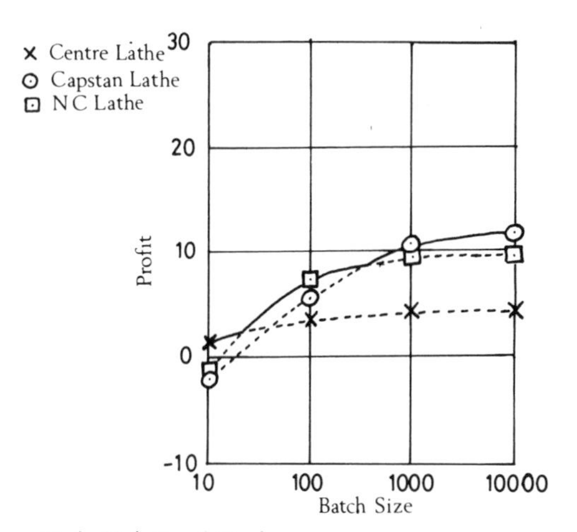

FIG. 7.19. Profit for various types of machine using a profit 'standard' of 31·6% for a centre lathe producing a batch of 10.

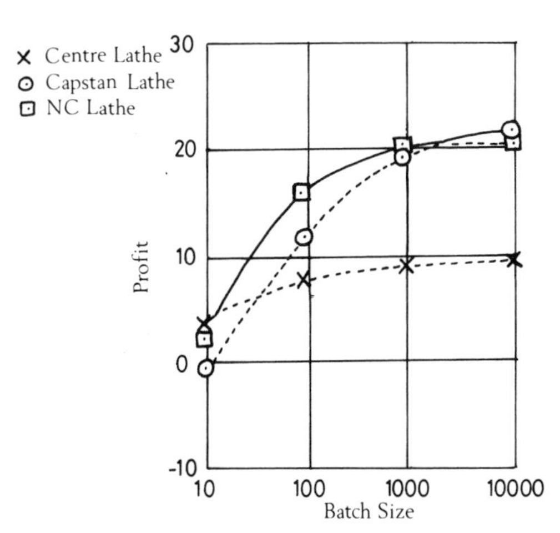

FIG. 7.20. Profit for various types of machine using a profit 'standard' of 100% for a centre lathe producing a batch of 10.

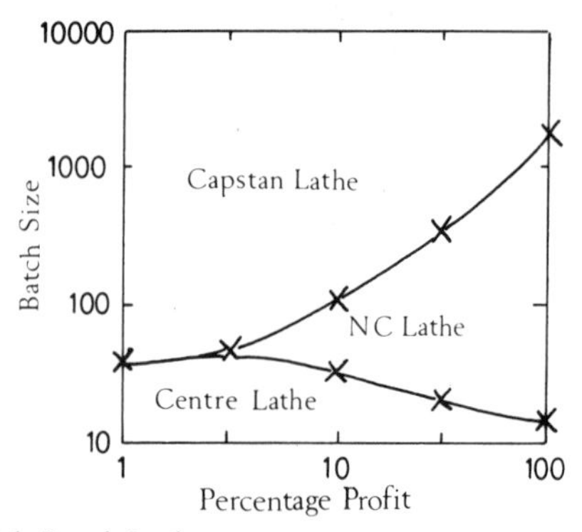

FIG. 7.21. Profit–batch size relationship based on centre lathe batch size 10.

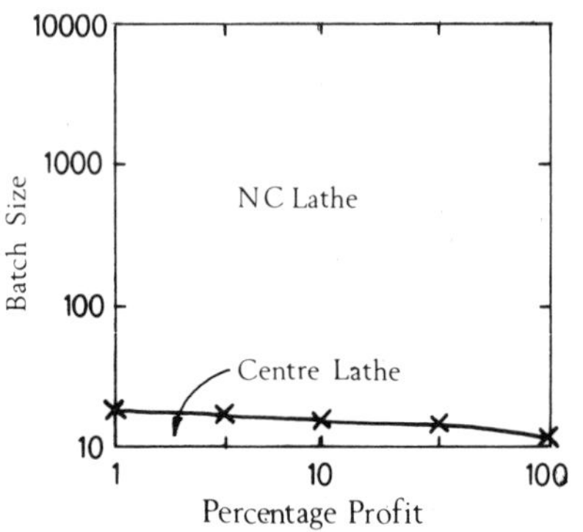

FIG. 7.22. Profit–batch size relationship based on centre lathe batch size 10.

information required to operate a maximum profit criterion is much more comprehensive than that required to operate a minimum cost or maximum production rate criterion and consequently is more susceptible to error. In practice, it is likely that a minimum cost criterion would be considered to be easier to apply and therefore more acceptable; further, as mentioned earlier, economic cutting conditions can be achieved regardless of the size of the batch. Best utilisation of machines is difficult to achieve irrespective of the criterion used and for realistically sized machine shops can only be effected by computer-aided production planning.

7.7. MACHINABILITY DATA APPLIED TO MILLING

With the exception of face milling, which in itself represents only a very small proportion of all types of milling, virtually all other types of milling cutter are manufactured from high-speed steel. Further, for a large proportion of these, the cutter is both an expensive item to purchase and to regrind when the cutting edge has been deemed to be fully worn and, for many conditions, it is time-consuming to change cutters. Thus, whereas with single point tools and particularly in the case of throw-away inserts both the cost of each cutting edge of the tool and the cost of changing the tool are small, for milling cutters both these costs are generally large. Examination of the equation (Eqn. 7.4) for economic cutting speed which is of the form:

$$V_c = C_1 \left(\frac{M}{(MT_I + C_T)(-k-1)} \right)^{-1/k}$$

and which would typically have values of, say, 10 min for T_I and, possibly, 1500 pence for C_T and taking $C_1 = 80$, $k = -5$ and $M = 5$, would give V_c as approximately 19 m min^{-1}. The equation for economic cutting speed for milling is somewhat different to that for turning but the essential features are the same and the general conclusion that can be drawn is that economic cutting speeds in milling are much lower than their equivalents in turning. If, as would be usual, cutting speeds even lower than the economic cutting speeds were to be utilised to increase tool life significantly without a disproportionately large increase in the cost of production, changes in machinability of the tool–workpiece combination would have a much smaller effect on performance in milling than in turning.

In the special, yet proportionately significant case of transfer line milling, where a variety of operations such as milling, drilling, tapping, reaming, etc. are carried out sequentially and where a single premature tool failure can result in a stoppage of the complete line, it is important that cutting conditions be chosen conservatively so that the incidence of premature tool failure is reduced. For this type of machine various possibilities exist for minimising the effect of premature tool failure such as automatic sensing of failure and tool replacement for simple tools such as drills, taps and reamers and the introduction of buffers at strategic points on the machine that ensure that breakdowns or planned tool replacements on one section will not stop production on other sections. However, even with these ‘regulators’, the general concept of extending tool life by the use of downgraded cutting conditions is still sound practice. From the previous evidence, it is reasonable to assume that for milling, improvements in machinability can be reflected in improved performance (lower cost production) but that this will not be as significant in milling as in turning. Nevertheless, a knowledge of the machinability characteristics is still very important since it is only from these that appropriate cutting conditions can be chosen. To date, even though much effort has gone into the generation of machinability data for milling, there are no accepted standards for these types of cutting tool and difficulties in formulating standards have arisen because of the variety of types of tool, the various ways in which the tools can be used, and the complex nature of the cutting process. As a consequence of this lack of standardisation (and for those cases where some data are available it is possible that the data are not appropriate for the specific operation being considered) it is probable that the user can, by careful monitoring of production, collect data which can be used meaningfully to obtain economic cutting conditions for a given production situation, i.e. a slow manual adaptive control strategy can be utilised.

7.8. RELIABILITY OF MACHINABILITY DATA

It has long been accepted that the results of tests to determine the constants k and C_1 of Taylor's equation ($VT^{-1/k}=C_1$) are generally subject to large errors and, typically, values for these constants for reasonable confidence would be $\pm 30\%$. The Production Engineering Research Association carried out 'controlled' tests whereby they specified the cutting conditions, the cutting tool and the workpiece material and asked several laboratories throughout Europe to carry out machinability tests to the ISO specification for turning. The results of these tests varied widely and correlation from laboratory to laboratory was very poor—this was probably the result of one of two factors: either the machine tools on which the tests were carried out had significantly different characteristics, or, more likely, there was a wide variation in composition and properties of the workpiece materials within the general specification. Work carried out by the authors and others indicates that even with the most stringent control of tool, workpiece, machine tool, and cutting conditions the results of wear tests are variable and even under 'ideal' conditions it would not be possible to obtain values of k and C_1 which would be accurate to better than $\pm 20\%$ for a catastrophic failure criterion applied to high-speed-steel tools or $\pm 15\%$ when applied to the flank or crater wear of cemented carbide tools. These figures may, at first sight, appear to be undesirably large but when considered in relation to the effect of cutting speed on tool life they do not represent oversignificant error in calculations of tool life, i.e. if the worst values of k and C_1 were considered the values obtained for cutting speed for a given tool life would not be appreciably different to those obtained using the mean values. If it is accepted that the cutting tools will have constant properties and that the varying characteristics from machine to machine will not significantly affect the machining characteristics then differences in machinability can be said to be *primarily* caused by changes in properties of the workpiece material within a grade. It could be argued that a user could specify and be able to purchase workpiece materials to a narrower specification than the broad grade specification but this would invariably mean that other users would have to accept wider variations within a grade. Generally, users are not too specific on workpiece specification and consequently variations in machinability of the order outlined above are almost inevitable and account has to be taken of this when selecting cutting conditions to meet a specific requirement of performance.

In conclusion, it can be said that even with doubtful data it is likely that, for most applications, the correct choice of machine can be made but the choice of cutting conditions will not be optimal whatever the criterion of performance chosen. However, it is more likely that the machine will not be capable of performing optimally due to either a low maximum rotational speed or lack of power at the spindle and, as a consequence, machinability data as determined from standard tests will generally be adequate.

REFERENCE

1. Cook, N. H. (1966). *Manufacturing analysis*, Addison Wesley Publishing Company, London.

Appendix 1

ANALYSIS TO DETERMINE CUTTING TEMPERATURES IN SINGLE POINT METAL CUTTING

The heat sources and heat sinks associated with a single point orthogonal metal cutting operation are shown in Fig. A1.1 where it can be seen that there are two heat sources, the primary deformation zone and the secondary deformation zone, and three heat sinks, the tool, the workpiece and the chip. It is assumed that no heat enters the atmosphere directly and whilst this is clearly not true, the amount involved is insignificant.

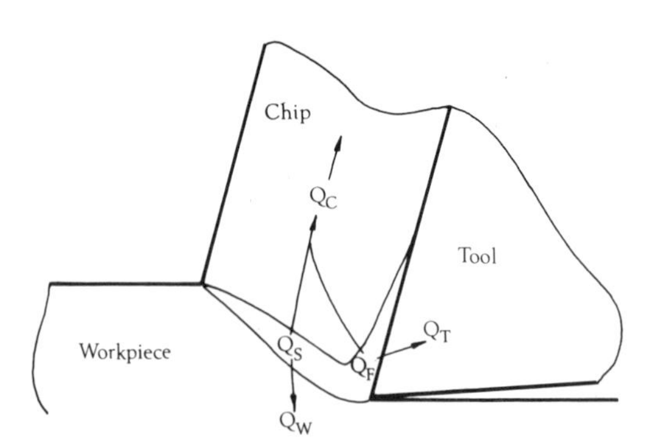

FIG. A1.1 Heat sources and sinks in orthogonal metal cutting.

If a heat balance is now carried out then the total heat per second input to the system, Q, is given by

$$Q = Q_s + Q_F \qquad [A1.1]$$

where Q_s is the heat generated in primary deformation per second, and

Q_F is the heat generated in secondary deformation per second. The total heat per second output from the system, Q, is given by

$$Q = Q_T + Q_W + Q_C \qquad [A1.2]$$

where Q_T is the heat input to the tool per second; Q_W is the heat input to the workpiece per second; and Q_C is the heat input to the chip per second.

If the small amount of heat carried away by the tool is now ignored then

$$\begin{aligned} Q &= Q_s + Q_F \\ &= Q_W + Q_C \end{aligned} \qquad [A1.3]$$

and since, effectively, all the work done in metal cutting is converted into heat, the total heat generated per second, Q, will be given by

$$Q = F_c V_c \qquad \text{(watts)} \qquad [A1.4]$$

where F_c is the tangential cutting force (N); and V_c is the cutting speed ($m\,s^{-1}$). (There will also be small amounts of heat generated due to work done in the thrust direction but this will be insignificant compared with that done in the cutting direction.) Further, since the heat generated per second due to secondary deformation, Q_F, is a result of overcoming friction on the tool face

$$Q_F = F V_F \qquad \text{(watts)} \qquad [A1.5]$$

where F is the friction force acting on the tool (N); and V_F is the chip velocity ($m\,s^{-1}$).

From an analysis of the force system acting on the tool

$$F = F_c \sin\alpha + F_T \cos\alpha \qquad [A1.6]$$

where F_c is the tangential cutting force; F_T is the normal cutting force; and α is the rake angle.

From volume flow considerations

$$V_F = V_c r_c \qquad [A1.7]$$

where r_c is the cutting ratio and is defined as the ratio of the undeformed chip thickness to the deformed chip thickness.

Thus, since $Q = Q_s + Q_F$

$$\begin{aligned} Q_s &= F_c V_c - F V_c r_c \\ &= V_c (F_c - F r_c) \end{aligned} \qquad [A1.8]$$

Of the heat generated in the primary deformation zone a portion is conducted into the workpiece whilst the remainder is conducted and transported into the chip. It has been shown that the heat conducted into the chip is negligible compared with that conducted into the workpiece and that transported into the chip. Using this assumption, an analysis carried out by Weiner (1) showed that for the boundary conditions indicated in Fig. A1.2, a unique relationship exists between the

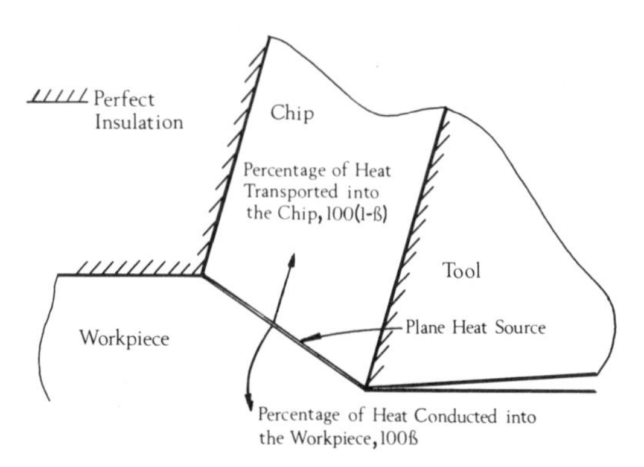

FIG. A1.2 Model used by Weiner to determine the percentage of primary deformation zone heat conducted into the workpiece.

percentage of heat conducted into the workpiece, the product of the thermal number, R_T, and the tangent of the shear angle, tan ϕ. This relationship is shown in Fig. A1.3 and subsequent experiments by several authors have indicated that this relationship is valid.

The thermal number, R_T, is given by

$$R_T = \frac{\rho c V_c t_1}{k} \quad [A1.9]$$

where ρ is the density of the workpiece; c is the specific heat of the workpiece; t_1 is the undeformed chip thickness; and k is the thermal conductivity of the workpiece.

Using appropriate values of ρ, c and k, the experimental values of V_c and t_1, and the determined value of ϕ [$\phi = \tan^{-1}(r_c \cos\alpha/(1 - r_c \sin\alpha))$], the proportion of heat conducted into the workpiece, β, can be determined. Since heat flow rate is the product of mass flow rate, specific heat and change in temperature, the average temperature rise in primary deformation, θ_s, of the material flowing to form the chip will be given by

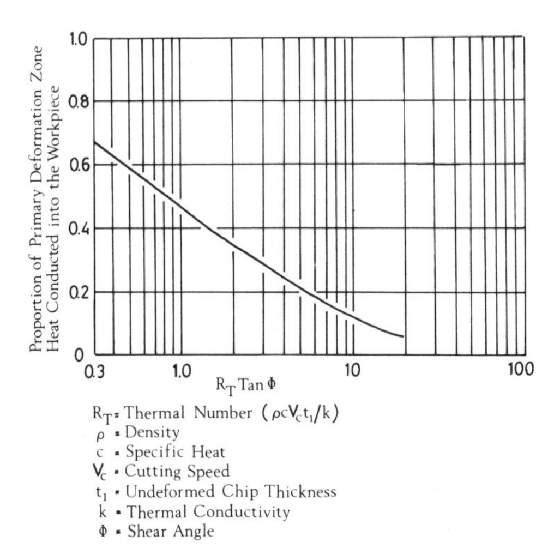

FIG. A1.3 Weiner's relationship for heat conducted into the workpiece.

$$\theta_s = \frac{Q_s(1-\beta)}{\rho c V_c t_1 a} \qquad [A1.10]$$

where a is the width of cut.

Using the assumption that a negligible amount of heat is conducted into the tool, all the heat generated in secondary deformation will be carried away by the chip and the average temperature rise of the chip as it passes through the secondary deformation zone, θ_F, will be given by

$$\theta_F = \frac{Q_F}{\rho c V_c t_1 a} \qquad [A1.11]$$

Work by Rapier (2) using the boundary conditions indicated in Fig. A1.4 led to the expression

$$\frac{\theta_m}{\theta_F} = 1{\cdot}13\sqrt{\frac{R_T}{\alpha_1}} \qquad [A1.12]$$

where θ_m is the maximum temperature rise in secondary deformation; and α_1 is the ratio of the contact length on the face to the deformed chip thickness.

Experimental work by Boothroyd (3) indicated that the maximum temperature rise in secondary deformation was generally considerably

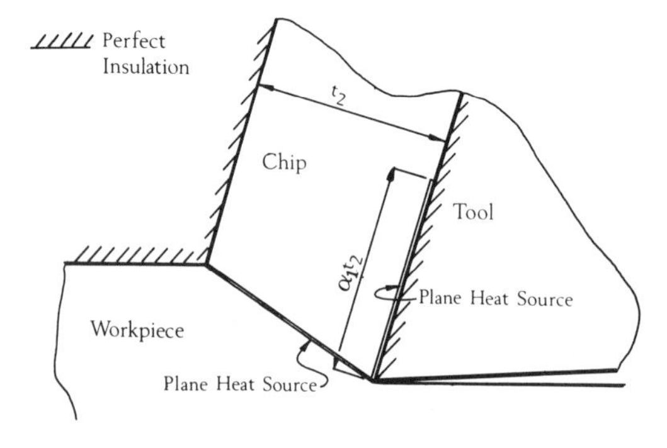

FIG. A1.4 Rapier's model to determine the maximum temperature rise of the chip due to secondary deformation.

less than that predicted by Rapier (2) and he concluded that the most probable source of error was due to Rapier's assumption of a plane heat source on the face of the tool. Accordingly, Boothroyd (3) developed a model where the heat source due to secondary deformation was rectangular (Fig. A1.5) and extended over the whole of the contact length. (It was known that a triangular heat source would be more realistic but it was felt, and later justified, that a rectangular source of the same area would give adequate results.) The ratio of the rectangular heat source width to the deformed chip thickness was defined as γ.

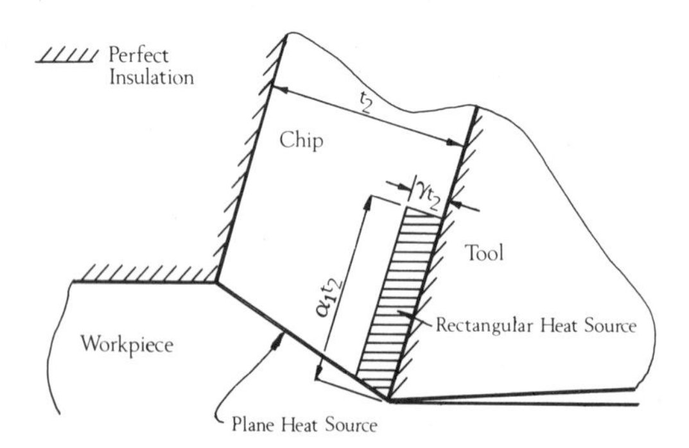

FIG. A1.5 Model used by Boothroyd to determine the maximum temperature rise of the chip due to secondary deformation.

Figure A1.6 shows the relationship between θ_m/θ_F and γ for a range of values of R_T/α_1 and, as could be expected if γ is set equal to zero (a plane heat source), the Rapier solution is obtained.

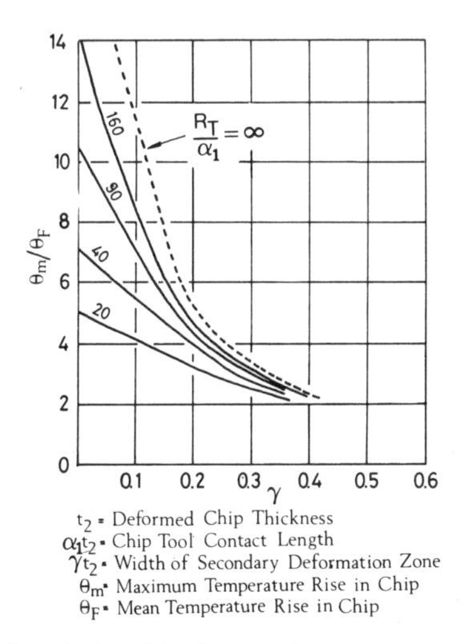

FIG. A1.6 Boothroyd's relationship for maximum temperature rise in the chip.

In practice, the contact length, α_1, can be determined relatively easily by experiment but the average width of the secondary deformation zone, γ_1, is more difficult to estimate. Results from tests on various materials and for a range of cutting conditions have indicated that γ is typically of the order of 0·2.

From the above, the maximum temperature of the chip, θ_{max}, is given by

$$\theta_{max} = \theta_s + \theta_m + \theta_0 \qquad [A1.13]$$

and the mean temperature of the chip is given by

$$\theta_{mean} = \theta_s + \theta_F + \theta_0 \qquad [A1.14]$$

where θ_0 is the ambient temperature.

Both θ_{min} and θ_{max} have been used by many researchers as the basis of temperature measurement for determining the relationship between cutting temperature and tool life. Further, although the analysis is basically for orthogonal cutting it would appear that it is equally valid for oblique cutting.

REFERENCES

1. WEINER, J. H. (1955). Shear plane temperature distribution in orthogonal cutting, *Trans. A.S.M.E.*, **77**, 1331.

2. RAPIER, A. C. (1954). A theoretical investigation of the temperature distribution in orthogonal cutting, *Brit. J. Appl. Phys.*, **5**, 400.
3. BOOTHROYD, G. (1963). Temperatures in orthogonal metal cutting, *Proc. I. Mech. E.*, **177**, 789.

Appendix 2

ANALYSES FOR TWO SHORT ABSOLUTE MACHINABILITY TESTS

A2.1. THE VARIABLE-RATE MACHINING TEST

The variable-rate machining test was developed by Heginbotham & Pandey (1) to produce accurate values for the constants k and C_1 of Taylor's tool life equation, $VT^{-1/k}=C_1$, without using as much time and material as the traditional long absolute machinability test. It was felt that for the test to be practical it should firstly be capable of being carried out using axial turning and secondly produce wear over the whole range of cutting speeds to be investigated. To achieve this, a test was developed where the workpiece material was axially turned with a constant rate of increase of cutting speed with cutting time. For this, and by reference to Fig. A2.1, it can be seen that if V_I is the initial cutting speed and V_F is the final cutting speed then the cutting speed after t minutes of cutting, V, will be given by

$$V=V_I+\frac{(V_F-V_I)t}{t_1} \qquad [A2.1]$$

where t_1 is the total cutting time for one 'pass'.

If the tool life to some chosen criterion is considered to be T minutes then in the small interval of time, dt, the portion of tool life used will be given by (dt)/T and if Taylor's equation, $VT^{-1/k}=C_1$, is valid then substituting for T will give

$$\frac{dt}{T}=\left(\frac{V}{C}\right)^{-k} dt \qquad [A2.2]$$

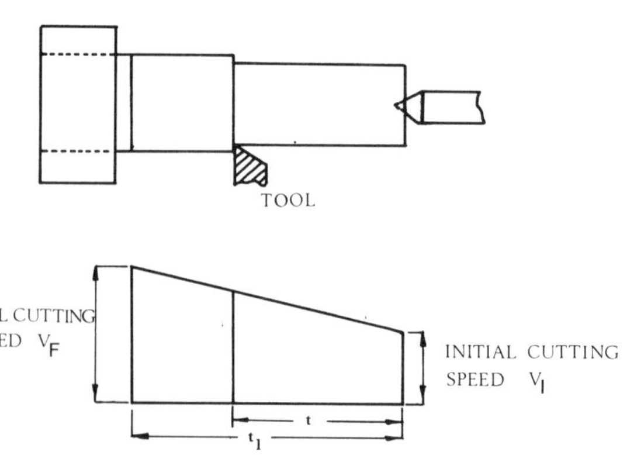

FIG. A2.1. Cutting speed–time relationship for the variable-rate machinability test. Cutting speed after time $t = V_I + \left(\frac{V_F - V_I}{t_i}\right)t$.

which, on substituting for V from Eqn. [A2.1] leads to

$$\frac{dt}{T} = \left\{\frac{1}{C_1}(V_I) + \frac{(V_F - V_I)t}{t_1}\right\}^{-k} dt \qquad [A2.3]$$

Putting $Z = V_F/V_I$ and integrating Eqn. [A2.3] between limits $t = 0$ and $t = t_1$ will give the total portion of tool life consumed per pass, X; i.e.

$$X = \left(\frac{V_I}{C_1}\right)^{-k} \int_{t=0}^{t=t_1} \left(1 + \frac{(Z-1)t}{t_1}\right)^{-k} dt$$

$$= \left(\frac{V_I}{C_1}\right)^{-k} \left[\frac{Z^{-k+1} - 1}{Z - 1}\right] \frac{t_1}{-k+1} \qquad [A2.4]$$

but

$$t_1 = \frac{(V_I + V_F)S}{2\pi DL} \qquad [A2.5]$$

where L is the cut length of the workpiece; D is the diameter of the workpiece; and S is the feed.

Substituting for Z in Eqn. [A2.5] gives

$$t_1 = \frac{V_I(1+Z)S}{2\pi DL}$$

and substituting this in Eqn. [A2.4] gives

$$X = \frac{\pi DL}{f(\mathrm{k}+1)V_{\mathrm{I}}}\left(\frac{V_{\mathrm{I}}^{-\mathrm{k}}}{\mathrm{C}_1}\right)\left[\frac{Z^{-\mathrm{k}+1}-1}{Z^2-1}\right] \qquad [A2.6]$$

For repeated testing under the same conditions

$$X = \frac{W_{\mathrm{m}}}{M_0 W_0} \qquad [A2.7]$$

where M_0 is the number of repeat passes; W_0 is the flank wear land at failure; and W_{m} is the flank wear land after M_0 passes.

Taking the simplest set of conditions where, for two different tests, the feeds and the ratio V_{I} to V_{F} are kept constant

$$-\mathrm{k} = 1 + \frac{\log \alpha}{\log \beta} \qquad [A2.8]$$

where $\alpha = \dfrac{X_1 D_2 L_2}{X_2 D_1 L_1}$, and $\beta = \dfrac{V_{\mathrm{I}_1}}{V_{\mathrm{I}_2}}$

Substituting the value of k obtained from Eqn. [A2.8] into Eqn. [A2.6] will give the value of the constant C_1.

Table A2.1 shows a comparison between the time taken and the amount of material used for a typical variable-rate test and a standard absolute machinability test.

TABLE A2.1
COMPARISON BETWEEN CONVENTIONAL AND VARIABLE-RATE MACHINABILITY TESTING

Test	*Mass of steel used (kg)*	*Testing Time (min)*	*No. of test tools*
Conventional	123	203	6
Variable-rate	18	35	2

A2.2. THE STEP TURNING TEST

As a result of the work reported on the variable-rate machinability test developed by Heginbotham & Pandey (1), Kiang & Barrow (2) proposed that variable-rate machining tests could be represented by a large number of constant speed steps which would, firstly, allow the tests to be carried out on a standard centre lathe and secondly, eliminate any possible temperature lagging effects which might occur with variable-rate machining. The method had been used previously (1) with only two cutting speeds but it was felt that this would be too restrictive and could not adequately predict the constants of Taylor's tool life equation.

To facilitate the analysis for the step turning test the concept of degree of wear, G, was introduced and this was defined as the ratio of the actual tool wear to the wear required to reach the desired criterion of tool failure. (This is synonymous with portion of tool life, X, used by Heginbotham & Pandey (1).)

Considering the Taylor tool life equation as being $VT^{-1/k}=C_1$, this may be rearranged as $T=(V/C_1)^k$ and the amount of tool life used per unit time at a particular speed will be given by

$$\frac{1}{T}=\left(\frac{C_1}{V}\right)^k \qquad [A2.9]$$

If it is now reasonably assumed that the flank wear–cutting time in the secondary wear zone is linear then, when the test is performed in a discrete manner, the tool life can be determined by the direct summation of the individual tool life values reached during each step of constant speed turning. Thus, when the tool is considered to have failed,

$$\sum_{i=1}^{i=m}\left(\frac{V_i}{C_1}\right)^{-k} t_i=1 \qquad [A2.10]$$

where m is the number of steps to failure; and t_i is the time of cutting at each speed V_i. Conversely, if after m stops the fraction of tool life consumed is G then

$$\sum_{i=1}^{i=m}\left(\frac{V_i}{C_1}\right)^{-k} t_i=G \qquad [A2.11]$$

If two tests are conducted then two simultaneous equations will result which, after expansion, may be written as

$$\left(\frac{V_{11}}{C_1}\right)^{-k} t_{11}+\left(\frac{V_{21}}{C_1}\right)^{-k} t_{21}+\ \ldots .\left(\frac{V_{m1}}{C_1}\right)^{-k} tm_1=G_1 \qquad [A2.12]$$

and

$$\left(\frac{V_{12}}{C_1}\right)^{-k} t_{12}+\left(\frac{V_{22}}{C_1}\right)^{-k} t_{22}+\ \ldots .\left(\frac{V_{m2}}{C_1}\right)^{-k} tm_2=G_2 \qquad [A2.13]$$

If the cutting conditions are now chosen such that

$$t_{11}=t_{21}=t_{31}\ \ldots\ =t_{m1}=t_1$$

$$t_{12}=t_{22}=t_{32}\ \ldots\ =t_{m2}=t_2$$

and

$$m_1=m_2=m$$

and that, further,

$$V_{11}=V_1,\ V_{21}=aV_1,\ V_{31}=a^2V_1\ \ldots\ V_{m1}=a^{m1-1}V_1$$

and

$$V_{12}=V_2,\ V_{22}=aV_2,\ V_{32}=a^2V_2\ \ldots\ V_{m2}=a^{m2-1}V_2$$

then dividing Eqn. [A2.12] by Eqn. [A2.13] will give

$$\frac{t_1}{t_2}\left(\frac{V_1}{V_2}\right)^{-k}\frac{\sum_{i=0}^{i=m-1}(a^k)^i}{\sum_{i=0}^{i=m-1}(b^k)^i}=\frac{G_1}{G_2}$$

Assuming that the log V–log T relationship of the Taylor equation is valid and putting $b=a$ gives

$$-k=\log\left(\frac{G_1t_2}{G_2t_1}\right)\Big/\log\left(\frac{V_1}{V_2}\right) \qquad [A2.14]$$

Once k has been determined from Eqn. [A2.14], C_1 can be calculated from either of Eqns. [A2.10] and [A2.11].

If the simplified equations are not used, values for k and C_1 can still be determined but a computer solution will be necessary.

Initial tests carried out by Kiang & Barrow (2) indicated that the correlation obtained between the values for k and C_1 using the test procedure and those obtained using a standard long test was not good. It was thought that this was attributable to the high wear rate in the primary wear region and this was in agreement with conclusions reached by Heginbotham & Pandey (1) after their testing. When the cutting tools were pre-worn to a level which extended beyond the primary wear stage and when the values for G were modified accordingly, agreement with the results of the standard long absolute machinability test was good.

Table A2.2 shows a comparison between the time taken and the amount of material used for a typical step turning test and a standard absolute machinability test.

TABLE A2.2
COMPARISON BETWEEN CONVENTIONAL AND STEP TURNING MACHINABILITY TESTING

Test	*Mass of steel used (kg)*	*Testing time (min)*	*No. of test tools*
Conventional	64	660	6
Step turning	15	137	4

REFERENCES

1. HEGINBOTHAM, W. B. & PANDEY, P. C. (1967). A variable rate machining test for tool life evaluation, *Proc. 8th Inst. M.T. D.R. Conference*, September, 163–71.
2. KIANG, T. S. & BARROW, G. (1971). Determination of tool life equations by step turning tests, *Proc. 12th Int. M.T.D.R. Conference*, September, 379–86.

INDEX